The Porpoise-Given Life

The Porpoise-Given Life

Inspiration from the Cetacean Nation

Chris B. Hughes

Providence House Publishers
WWW.PROVIDENCEHOUSE.COM
FRANKLIN, TENNESSEE

Printed in the United States of America

12 11 10 09 08 1 2 3 4 5

Library of Congress Control Number: 2008930601

ISBN: 978-1-57736-413-9

Cover and page design by Joey McNair

PROVIDENCE HOUSE PUBLISHERS
238 Seaboard Lane • Franklin, Tennessee 37067
www.providencehouse.com
800-321-5692

Contents

Preface

Why I Did It and What's Inside

There are many reasons, some just for fun and some more serious, why I wrote this book and offer it to you.

I am an incurable punster. Nothing is sacred if I'm in the right (or terribly wrong) mood. Take my twist on the popular praise song, "Lord I Lift Your Name on High." My version of the chorus changes "my debt to pay" to "my dead toupee." As a child, I drove my big sister to tears by changing our painstakingly prepared rendition of the hymn "There is a Balm in Gilead" to "There is a Bomb in Gilead," with appropriate sound effects added at the last minute (and ahead of its time) as we sang for our parents.

I've been a lifelong fan of Weird Al Yankovic, the great master of pop parody. Given that, and my preoccupation with Jesus and the Gospel, it was inevitable that I would turn Sheryl Crow's "Soak Up the Sun" into "Soak Up the Son," a celebration of Christian discipleship[1]; Vanilla Ice's "Ice, Ice, Baby" into "Christ, Christ, Baby" (a little advent ditty, "Word to your Mother"); and Olivia Newton John's "Let's Get Physical" into "Let's Get Biblical." I've been fiddling with John Mayer's "Your Body is a Wonderland" as a Communion hymn.

I was in the project stage of my doctor of ministry degree program when I first encountered *The Purpose-Driven Life* by Rick Warren.[2] I was deeply immersed in the search for new metaphors for being and doing church in the twenty-first century. So, having grown up on the Gulf Coast of Florida with a love shared by many on the planet for dolphins and porpoises, I couldn't resist punning around with Warren's book title and entertaining the idea of a treatise on the life lessons given to us by those delightful mammals who live in the sea. Admittedly, the analogies and metaphors drawn here are mostly from dolphin studies and stories, but the pun doesn't work with "dolphin." Porpoises, dolphins, and whales are among the species in the order Cetacea, hence the subtitle, "Inspiration from the Cetacean Nation."

I found two works that both play with Warren's book title and offer perspectives on living that challenge that of *The Purpose-Driven Life*. Robert M. Price's book, *The Reason-Driven Life*,[3] is an often scathing, entertaining, and insightful non-theist full frontal rebuttal of Warren's book. *The Porpoise Diving Life* by William S. Dahl offers chapters 41–80, "picking up where the purpose-driven perspective being peddled [by Warren] peters out."[4] Dahl's e-book is for those who don't want to give in to the idea that all of life is predetermined and that all experiences of life are set up by God to test or reward them.

I'm not responding to Warren's version of Calvinistic evangelical Christian dogma by offering an opposing system of dogma or a chapter-by-chapter riposte. (Although I will take a stab in another book at constructing a post-Calvinist, postmodern evangelical theology for the twenty-first century as an alternative to the seventeenth-century deterministic theology so many people cling to as if it were handed down from Mount Sinai with the Big Ten).

Millions of people, possibly including you, have found comfort, inspiration, and reassurance in *The Purpose-Driven Life*. There is much in Warren's book I found helpful, and much that I found disturbing, paternalistic, and even cruel. Like you, I need some structure, some handles, some participation in a vision larger than my own. I want to be part of the answer to the prayers I pray for others. I want to be a part of co-creating God's preferred vision for the world. I want to be a partner with you and God so that earth is more heavenly in this life, not merely the next.

I want to offer a more playful, constructive, and choice-driven response to life. I wrote this book because life is about more than being "driven." Jesus invites us to abundant life, encouraging us to, "Take my yoke upon you, and learn from me; for I am gentle and lowly in heart, and you will find rest for your souls. For my yoke is easy, and my burden is light" (Matt. 11: 29–30). He invites us to get in harness with him, to join him in his ongoing mission to bring good news to the poor; feed the hungry; welcome the stranger; make sure everyone on the planet has clean water, safe shelter, and adequate clothing; and share one another's burdens. Duffey Robbins, popular speaker and professor of youth ministry at Eastern College, told of his confusion as a young teenager when he thought he heard the preacher urge the congregation to be "constipated for the Lord." Of course, the preacher had said, "be consecrated for the Lord" (as in set aside, prepared, and blessed for service, witness, and mission).

Robbins thought to himself, "Why would anyone want to be constipated for the Lord?" As he looked around the sanctuary and saw the pinched, pained, and strained faces of so many of the worshippers, he realized that there certainly are a lot of constipated Christians whose restrictive faith rates about a 9.5 on the "sphincter" scale.

I want to be part of a new movement—as my friend, storyteller, musician, and folk theologian Ed Kilbourne puts it—a movement that unbinds and releases us for free and passionate living.

I wrote this book because life is not a dress rehearsal for some future eternity. This life we find ourselves in *right now* is part of eternity. David Olive, president of Bluefield College in Bluefield, Virginia, suggests that the opposite of fear is faith. I want to move, and invite you to move with me, away from a pat, practiced, and fear-driven rightness as a hedge bet against eternal damnation toward a fearless and passionate gracefulness as a sure bet on purpose-fullness in the here and now.

And honestly, I wanted to get this out before the pun didn't work anymore because people don't remember Warren's book . . . as if that could happen, given the constant stream of *Purpose-Driven* spin-off products. Thankfully, Warren has been fine-tuning his life's purpose to include being a vocal leader in expanding the faith-based public policy agenda of evangelical Christians beyond gay marriage and abortion to include biblically mandated attention to stewardship of the earth, global poverty and hunger, genocide, and a consistent ethic of life.

I'm indebted to the people who have given their lives in the vocation of learning about and learning from porpoises, whales, and dolphins. Heathcote Williams's classic and moving poetic work, *Whale Nation*,[5] was my single best source for insight and quotable quotes about cetaceans. *Souls in the Sea* by Scott Taylor, founder of the Australia-based Cetacean Studies Institute, provides a sweeping history of the dolphin's hold on the imagination of humankind.[6]

I know that there are many, perhaps even most, on this planet who spend all of their energy on simple survival. I also know that many of them know the simple joy of simply living. It is not my intention to minimize their struggle for life with my invitations to joyful, cooperative, and interdependent living. In fact, I am moved now more than ever to be in solidarity with the "have nots" in a world of plenty.

I could not have sustained the effort to finish this project without the constant support and encouragement of my wife, Gloria. I hope I do justice to all who contribute to my delight-full life: my family and friends, my extended God-family, my teachers and mentors (some I know "in person" and some only through books), and the people who continue to invite me to walk with them in spiritual growth experiences.

There are more complete studies and celebrations of the wonders of cetaceans out there.[7] I'm just a fan. There are more astute theological works out there.[8] I'm not pretending to be a systematic theologian. I simply want to share some of my passion for life with the people I encounter in worship services, retreats, and training events, and in public places like stores, parks, and beaches—people like you, who may be tired as I am of heavy-handed airtight answers to some tough and elusive questions.

I am grateful for friend and mentor Dr. Leonard Sweet for his encouragement and wisdom.[9] He has a way of taking a pregnant metaphor and spinning a book around it. Like his *Gospel According to Starbucks,*[10] in which he mines the metaphor of coffee to encourage his readers to live and offer others an EPIC (Experiential, Participatory, Image-based, and Connective) experience of the gospel. I wanted to try my hand at metaphoraging with porpoises, dolphins, and whales as my central image.

What's Inside

The introduction celebrates the persistent images of porpoises, dolphins, and whales in the imaginations of humankind. Chapter 1 is an invitation to open your heart and mind to the wisdom of God revealed in the Creation, especially in the biblical similes and metaphors related to animals. (If you don't think you can learn anything from observing animals, you may want to skip this whole book—and much of the Bible). Chapter 2 deals specifically with the nature of whales, dolphins, and porpoises. You'll get a general overview of cetacean intelligence, physiology, and behavior that will serve as a foundation for the insights and perspectives of the rest of the book. There are a lot of notes in chapter 2. I cite a school of experts and "dol-fans" before stepping out with my own musings.

Each of the remaining chapters focuses on one aspect of life that is amplified and celebrated in the world of cetaceans. At the end of each chapter is a section called The Reflecting Pool. It contains some quotations to ponder, some questions to consider, and some suggested actions to take or experiences to explore. There is a passage from Judeo-Christian Scriptures to meditate on and a simple, one-sentence prayer that can be said in one breath and repeated as a mantra. The last chapter (yes, it's placed there and spelled "forward" on purpose), is a meditation on being more fully what we are created to be.

Unlike instructions in *The Purpose-Driven Life*, the book upon whose title I shamelessly play, I will not tell you how or in what order you should read *The Porpoise-Given Life*. Look at the table of contents. Thumb through the pages. Read what strikes you for as long as it holds your attention. Look at The Reflecting Pool sections and see what's there for you. I hope you will take this little book in the spirit it was written and offered to you: playfully, prayerfully, porpoise-fully.

Introduction

Lifeguards and Playmates

Three enduring images of cetaceans in the global imagination of the human race are: (1) whales, dolphins, and porpoises leaping from the sea in an explosive celebration of life; (2) dolphins and porpoises playing alongside and in the wakes of our boats; and (3) the astonishing role of the dolphin as a lifeguard and life giver to humans who find themselves literally over their heads in the home of their sea-living savior.

SCOTLAND—*DAILY RECORD,* AUGUST 30, 2000
A FRIENDLY DOLPHIN HAS SAVED A TEENAGE BOY FROM DROWNING.

Non-swimmer Davide Ceci, 14, was within minutes of death when dolphin Filippo came to his rescue. The friendly 61-stone creature has been a popular tourist attraction off Manfredonia in southeast Italy for two years. But now he is a local hero after saving Davide from the Adriatic when he fell from his father's boat. While Emanuele Ceci was still unaware his son had fallen into the waves, Filippo was pushing him up out of the water to safety. Davide said: "When I realised it was Filippo pushing me, I grabbed on to him." The dolphin bore down on the boat and got close enough for Davide's father to grab his gasping son. Davide's mother Signora Ceci said: "It is a hero, it seems impossible an animal could have done something like that, to feel the instinct to save a human life." Filippo has lived in the waters off Manfredonia since he became separated from a visiting school of dolphins. Maritime researcher Dr. Giovanna Barbieri said: "Filippo seems not to have the slightest fear of humans. I'm not surprised he should have done such a wonderful thing as to save a human."[1]

JAKARTA, INDONESIA (AS REPORTED IN THE *MIDLAND REPORTER/TELEGRAM*, MIDLAND, TX, JULY 18, 1982)

A Dutch helicopter pilot adrift in the Java Sea for nine days says a dolphin saved his life by nudging him to shore. Quentijn Fikke, 35, was flying from Jakarta to Balikpapan, East Borneo, when bad weather forced him into the sea. Fikke said the dolphin appeared during his second day in the water. "He stayed at my side more than eight days, only leaving me alone at nights but coming back in the morning. Each time he pushed me in the same direction." The dolphin disappeared after Fikke reached the beach. "There is no doubt that I owe my life to him."[2]

Humans have a historic and ongoing love affair with porpoises, dolphins, and whales. Stories of cetaceans saving people from drowning or drifting at sea are part the oral and written lore of most ancient and contemporary cultures around the globe. The image of the life-saving and life-giving dolphin is persistent in the imagination of man.

In 1945, a drowning woman reported being shoved to shore by what she identified as a porpoise as she neared unconsciousness after drifting into a strong undertow. She remembered:

> Someone gave me a tremendous shove and I landed on the beach, face down, too exhausted to turn over . . . when I did, no one was near, but in the water, about 18 feet out, a porpoise was leaping around, and a few feet beyond him, another large fish was leaping.

According to eyewitnesses, the "porpoise" was a dolphin and the other large fish was probably a fishtail shark![3]

On February 29, 1960, Mrs. Yvonne Smith of Stuart, Florida, fell from a boat off the coast of Grand Bahama Island. She described the purposeful guidance of a porpoise savior as it both protected her from sharks and barracuda and nudged her persistently toward land and safety. Yvonne describes this parting scene as her feet found the sea bottom at last: "As I turned toward shore, stumbling, losing [my] balance, and saying a prayer of thanks, my rescuer took off like a streak on down the channel."[4]

These stories include testimonies describing porpoises and dolphins keeping sharks at bay by encircling endangered swimmers, beating their tails against the surface of the water, and even ramming threatening predators. In June 2001, a group of fishermen from South Carolina drifted into the Gulf Stream after their boat sank some thirty-five miles off the coast of Georgetown. The struggling swimmers found themselves surrounded by a large group of mako, hammerhead, and tiger sharks. The fishermen were battered and scraped by the circling menace before a pod of several dolphins arrived and drove away the sharks. The dolphins stayed with the men all night and throughout the next day, fending off sea predators until the men could be rescued.[5]

An Ancient Account

Herodotus, historian of the ancient world, documented the first recorded story of a man being rescued by a dolphin in the fifth century BC. In the story, Arion, a renowned musician and poet, sails to the island of Sicily from his hometown of Lesbos in the northeastern Aegean Sea for a prestigious musical competition. Musicians from all over the known world would compete for wealth and fame (a kind of "Mediterranean Idol"). Gifted as he was on the kithara, a stringed instrument similar to the lyre, Arion wins first place, mounds of money, and a few laurel wreaths, and hires a ship and crew to take him and his riches to the court of King Periander in Corinth. However, when they reach the open seas, the greedy sailors plot to steal Arion's fortune and kill him.

Arion offers his riches in return for his life, but the bloodthirsty pirates want him dead. They offer Arion the choice of suicide by stabbing (and burial on land) or by jumping overboard. He was definitely caught between "Scylla and Charybdis" (which is the ancient Greek version of "between a rock and a hard place").

As a last request, Arion asks to sing a song and play his kithara on the ship's deck. Delighted to hear the best singer in the world, the murderous thieves agree. As the sweet sounds of Arion's playing and singing fills the air, a group of music-loving dolphins appears, frolicking alongside the ship. Arion sees his opportunity, praises the gods, and jumps overboard with his kithara in hand.

Presuming him dead, the pirates continue to sail on to Corinth. In the meantime, Arion rides on the back of one of the dolphins from the Sicilian Sea to Cape Tenaron, where he hires another ship to take him to Corinth.

Arriving ahead of the pirates, Arion tells the tale to King Periander, who is skeptical. The king hides Arion (in one version of the story, the king puts Arion in jail) and devises a scheme to expose the pirates when they arrive in Corinth with their ill-gotten wealth.

In another version of the story, the dolphin hero takes Arion all the way back to Corinth. Arion fails to push his rescuer back into deeper water and the dolphin dies. In response to the dolphin's heroic and unselfish act, King Periander ordered it to be buried, and a monument to be raised in its honor. To further honor Arion and his dolphin savior, the Greek god Apollo placed them among the stars as the constellation Delphinus (Latin for dolphin).

The dolphin was held in high esteem in ancient civilizations because it was believed that they were once humans; in some cultures, killing a dolphin was punishable by death.

Instinctive Reflex or Reflective Response?

Do dolphins actually respond, making a conscious decision to help humans in need, or are these merciful cetaceans just reacting out of instinct or even a sense of play? The helpful dolphins may be protecting endangered swimmers in the same way that they protect their young and members of their pod or family. Dolphins help dying members of their group to the surface so they can breathe. Like geese in migration, cetaceans will stay with one that is sick or disabled until they can rejoin the group. Cetaceans protect their pods from attacking sharks. They may just be extending to humans the care they give their kin.

Research over the last fifty years suggests that there may be some intention on the part of cetaceans to interact with humans. Cetaceans are very social and intelligent creatures that appear to be intentional in their behavior. They may be responding to another sentient creature in need, recognizing vulnerability, fear, and helplessness. Consider the news stories and personal testimonies. If the dolphins are just playing, why do they invariably steer struggling swimmers toward the shore? In less dire conditions, they certainly appear to initiate contact with humans for purposes of sheer pleasure. They may just do what they do on porpoise (I couldn't resist)! Whatever the explanation, it is reasonable for human beings who would otherwise be dead in the water but for the intervention of a dolphin to credit the sea mammal with saving their lives!

The Value and Limits of Metaphors

At a conference on pastoral leadership in the "new world" shaped by postmodernism,[6] Dr. Leonard Sweet observed, "In the battle for the hearts and minds of people, the one with the best metaphors wins. We are in a world where the making of culture depends on metaphors."[7] We use metaphors to help us name the unnameable and describe in part what is difficult to describe. Metaphors fill in the gaps of intuition and beauty left by mere definitions. Like extended analogies, metaphors give us a glimpse into the fullness of an idea, like the nature of God or what it means to be more fully human.

It's helpful to remember that the metaphor is not what it describes or attempts to name. God, for example, is none of the metaphors and images used to describe God. And yet, taken together, all of our metaphors help us get an emerging picture of the fullness of God. God is not a man or a woman. God is not a father or mother in the conventional sense, yet Jesus calls God "Father." Jesus refers to himself as a mother hen (Luke 13:34). God is not a king, or a general, or a shepherd. But each image gives us clues to our perception of God's self-revelation. We give God human characteristics so we can relate to God. In Jesus, we have the great, living metaphor of God, the Incarnation.

Of course, all metaphors have their limits. If you push on any metaphor hard enough it will break down at some point. There are exceptions to most any ideal. Realist and myth buster Philip Zaleski refers to a July 6, 1999, article in the *New York Times* that reports incidences of dolphins killing porpoises and even their own infants. The article suggests that dolphins don't even like humans (understandable given our historic treatment of their cousins, the whales, and our refusal to protect them from tuna nets and plastic six-pack rings) and that we romantically mistake instinctive play for purposeful rescue. After all, animals are animals.

Given that, the shocking occurrence of infanticide or out-of-character violence among cetaceans can almost always be explained by a parasite in the brain, some other mental or physical illness, or a threatening change in their ecosystem. As Zaleski points out, "Animals, unless diseased mentally or physically, simply do not engage in wanton murder. Only human beings have that honor."[8]

> The intelligent, playful, and caring qualities inherent to many aspects of bottlenose dolphin behaviour can at times resemble human behaviour at its best. This fact has undoubtedly contributed to their considerable universal appeal. But we now know that the behavior of bottlenose dolphins can also resemble humans at their [worst]. The one animal that we believed would always be totally "innocent," and trusted with our idealistic virtues of peace and well-being, has now shown us that it is just as capable of disturbing behaviour as humans.[9]

As we learn to do with human role models who we eventually discover are less than perfect, we will lift up the best that our friends the cetaceans can give, concentrating on their better nature for the benefit of ours. I'm choosing to move beyond cynicism and literal thinking as I look to the porpoise and the dolphin for inspiration and insight into the abundant life for which we are created and to which Jesus invites us.

We are prone to idealize cetaceans and make them over in our image rather than simply letting them be what they are and learning from them. Still, it would be a high compliment if others would say of us who confess that we are persons of faith and disciples of Jesus what Ernest Hemingway said of porpoises in *The Old Man and the Sea*, "They are good. They play, and make jokes, and love one another. . . ."[10]

> If an animal does something, we call it instinct; if we do the same thing for the same reason, we call it intelligence.
>
> *Will Cuppy,*
> *American humorist and journalist*

Chapter 1

An Invitation

For what can be known about God is plain to them, because God has shown it to them. Ever since the creation of the world his invisible nature, namely, his eternal power and deity, has been clearly perceived in the things that have been made.
Romans 1:19–20

Every animal knows more than you do.
Native American Proverb (Nez Perce)

The Judeo-Christian Scriptures, indeed, the sacred writings and wisdom stories of all the world's life-affirming faith traditions and the folklore of most indigenous peoples, are full of invitations to observe the natural world and all of its creatures for insight and wisdom. Parables about God's preferred vision for the world and analogies to God's nature based on the habits of familiar animals, and even on minerals and plants, abound in the words of the prophets, poets, and Jesus.

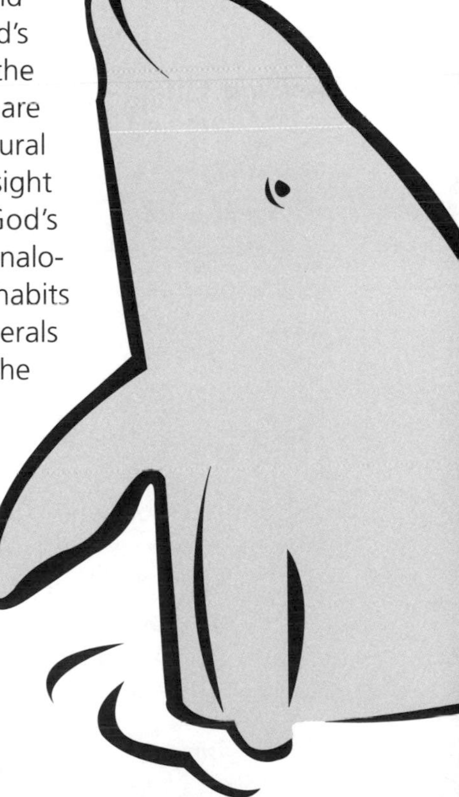

The psalmist declares in Psalm 19:1–2, "The heavens are telling the glory of God; and the firmament proclaims his handiwork." God is not in hiding. The image of God is upon all that exists. God's fingerprints, blueprints, and imprints are everywhere apparent. If we will pay attention as scientists, native peoples, and poets do, nature will teach us about life and the nature of God.

Eugene Peterson translates verse 2 like this: "Madame Day holds classes every morning, Professor Night lectures each evening" (THE MESSAGE). By attending Creation's classes, we are discovering that even in seeming chaos, there is order. We are discovering, with increasing wonder, the elegance of DNA and the mechanics of disease.

Jesus declares that even the stones cry out the glory of God, even if we are silent (Luke 19:40). Within these stones we have unlocked the inner power of atoms, for better and for worse. We are able to better cooperate with the natural order and move interdependently to ensure the future of the planet.

Jesus encourages us to look to presumably nonsentient parts of the natural order (such as rocks, streams, plants, and planets) and to other creatures (such as the mammals, fish, reptiles, and birds) for guidance, comfort, wisdom, and inspiration in our aspiration.

Carefree Living

> Therefore I tell you, do not be anxious about your life, what you shall eat or what you shall drink, nor about your body, what you shall put on. Is not life more than food, and the body more than clothing? Look at the birds of the air: they neither sow nor reap nor gather into barns, and yet your heavenly Father feeds them. (Matt. 6:25–26)

Peterson says that the birds are "carefree in the care of God" (Luke 12:22 THE MESSAGE). This doesn't mean that we don't need to eat and see to the feeding of others, or that we don't need clean water and we don't need to ensure safe drinking water for others. It does mean that people of faith will not live to eat, over-spending on processed foods and fancy meals; but instead will eat simply to live. It does mean that we need not spend billions of dollars per year on bottled water[1] (much of which is simply the

Animals have a great advantage over human beings: they never hear the clock strike, however intelligent they may be. They die without having any notion of death: they have no theologians to instruct them on the Four Ends of animals. Their last moments are not disturbed by unwelcome and often objectionable ceremonies. It costs them nothing to be buried. No one goes to law over their wills.

Voltaire,
Letter to the Compte de Schomberg

same quality as tap water available in most USAmerican communities); we should divert that money to simple water delivery systems in communities worldwide where children die from diarrhea and other preventable effects of dirty water. We can live and eat simply so others may simply live.

> Consider the lilies of the field, how they grow; they neither toil nor spin; yet I tell you, even Solomon in all his glory was not clothed like one of these. (Matt. 6:28–29)

You are more than clothing. Yes, most of us need to work to buy clothing. And we need to see that others are clothed. But we need not obsess over fashion. Like the creatures of the air, we are beautiful in our own right.

> Are not two sparrows sold for a penny? And not one of them will fall to the ground without your Father's will. But even the hairs of your head are all numbered. (Matt. 6:28–29)

Every creature, every action, is connected to every other. You may have heard of the theory postulated by Edward Lorenz known as the Butterfly Effect. It describes the dramatic effect of small variations introduced at the beginning of a larger process that ultimately affect the outcome. Like the flapping of a butterfly's wings (or the absence of flapping) at a particular place and time on one side of the world that contributes to the eventual formation of a hurricane (or not) on the other side![2]

Don't take from this passage the simplistic and self-serving idea that God loves people more than animals. Remember, God so loves *the world* (John 3:16). Learn from God's care for the fate of a sparrow that God cares about you and all creatures. Learn from God's response to one bird falling to the ground that we could surely take more care in our stewardship of the earth and all its creatures. Learn from the place of the sparrow in the grand scheme of God's unfolding creation that all of our actions matter.

Spreading Our Wings

> Jerusalem, Jerusalem, killer of prophets, abuser of the messengers of God! How often I've longed to gather your children, gather your children like a hen, Her brood safe under her wings—but you refused and turned away! (Luke 13:34 The Message)

Jesus echoes the psalmist and Isaiah in his lament over the stubborn children of God. In this passage we see a glimpse of God as protective mother, desiring to shelter her children and keep them from harm. But our willfulness, both a gift and a curse, keeps us from heeding the wisdom of those who raise us. The image of the mother hen calling out to her young also moves us to care for the unprotected among us, not so much against the wrath of God, but against the cold-heartedness of some people and the challenges of life.

> They who wait for the Lord shall renew their strength, they shall mount up with wings like eagles, they shall run and not be weary, they shall walk and not faint. (Isa. 40:31)

Those who don't try to run ahead of God, who humble themselves, slow down, and tune in to the rhythms of God and creation, get fresh strength. As Peterson translates it, "They spread their wings and soar like eagles" (THE MESSAGE). Those who learn to be still can walk with God, neither running ahead nor lagging behind. In the company of God, there is joy in the journey.

The eagle is the most mentioned bird in the Bible, cited more than two dozen times as an illustration of some desirable trait to be emulated by us or as a metaphor for God and God's disposition toward us.[3] Consider Psalm 103:2–5:

> Bless the Lord, O my soul, and forget not all his benefits, who forgives all your iniquity, who heals all your diseases, who redeems your life from the Pit, who crowns you with mercy and compassion, who satisfies you with good as long as you live so that your youth is renewed like the eagle's.

The eagle undergoes a dramatic change of life as it experiences the process of molting as an adult. During this time of molting, the noble and powerful eagle loses its feathers, and with them the strength and the ability to fly. Their eyesight, normally

> Three things are too wonderful for me; four I do not understand: the way of an eagle in the sky, the way of a serpent on a rock, the way of a ship on the high seas, and the way of a man with a maiden.
>
> *Proverbs 30:18–19*

so keen and reliable, fades. Unable to hunt, and preferring fresh kill, the molting eagle loses strength.

This is a profound time of choice for the aging eagle. Some choose to take the nourishment brought by other healthy and vital eagles in the midst of this degrading, depressing life cycle so that they may literally rise again with new feathers, renewed strength, and restored eyesight. Some simply roll over and die.

Many a human in midlife or beyond, having lost a life partner or the ability to do what they once could, is faced with this same kind of choice. When circumstances intervene to dramatically change our capacities, we can choose to re-imagine life, re-invest in relationships, and dream new dreams—or we can roll over and die. After all, Job laments, "My time is short—what's left of my life races off too fast for me to even glimpse the good. My life is going fast, like a ship under full sail, like an eagle plummeting to its prey" (Job 9:25–26 The Message). The swiftness of the eagle's flight serves as a metaphor for the brevity of life. So, make the most of it!

Snakes and Sheep

> Now the serpent was more subtle than any other wild creature that the Lord God had made. (Gen. 3:1)

In many cultures, the serpent (or snake) is a metaphor for slithering evil and smarmy deceit. Despite the valuable place of snakes in their many habitats and species in the planet's ecosystem, the serpent's fear factor looms large in the imagination. Jesus calls hypocritical religious leaders snakes and a viper's brood (Matt. 23:33) for demanding more from their flocks than they are willing to do themselves.

Peterson translates the word "subtle" as "clever" (Gen. 3:1 The Message). Remember how the serpent misrepresents God's motives and uses misguided human desire to confuse and convince Eve to go against God's explicit instructions?

> Lots of people talk to animals. . . . Not very many listen, though. . . . That's the problem.
>
> *Benjamin Hoff,* The Tao of Pooh

"Do I understand that God told you not to eat from any tree in the garden?" The Woman said to the serpent, "Not at all. We can eat from the trees in the garden. It's only about the tree in the middle of the garden that God said, 'Don't eat from it; don't even touch it or you'll die.'" The serpent told the Woman, "You won't die. God knows that the moment you eat from that tree, you'll see what's really going on. You'll be just like God, knowing everything, ranging all the way from good to evil." (Gen. 3:1–5 THE MESSAGE)

Eve thought it would be a good thing, even a God thing, to have all that knowledge. And it was such a beautiful, delicious-looking fruit. Perhaps the greatest sin of humankind is rationalization the ability to call a good thing bad and a bad thing good. Yes, we can be very clever and self-serving in our rationalizing, as we twist things around to justify our desires. Ends begin to justify means. Wants turn into needs.

When God shows up in the Garden of Eden looking for his beloved Adam and Eve, they hide. Yes, their eyes were open and they saw that they were naked. Of course, they had always been naked. Only now, in their newly enlightened state, they were ashamed of it. God asks them, "Who told you that you were naked?" (Gen. 3:11). A follow-up question may well have been, "And why did you listen?" We need reliable voices in our lives, not voices that lead us into despair and self-abuse, or callousness disguised as enlightened self-interest.

In stark contrast to the wily serpent bent on undermining the wonderful will of God for the creation and the destruction of humankind, the voice of the Good Shepherd is reliable. Jesus, the Good Shepherd, has only the good of his flock in mind. In John 10:11–27, Jesus explains:

> I am the Good Shepherd. The Good Shepherd puts the sheep before himself, sacrifices himself if necessary. . . . I know my own sheep and my own sheep know me. . . . The Father who put them under my care is so much greater than the Destroyer and Thief. (The Message)

The sheep are not stupid, just prone to wander. They know how to distinguish between the many voices playing on their fears and appealing to their self-interest to pick out and follow the voice of the Good Shepherd.

> An animal's eyes have the power to speak a great language.
>
> *Martin Buber*

First Peter 2:25 puts it this way, "You were lost sheep with no idea who you were or where you were going. Now you're named and kept for good by the Shepherd of your souls" (The Message). Jesus shows us the way to nourishment, abundant life, and servant leadership, caring for others in our care above ourselves. In both word and deed Jesus teaches us that, "No greater love has a person than this, than to lay down one's life for a friend"—or a neighbor (defined as anyone in need), a fellow sharer of the planet, all our kin, and our brothers and sisters, since we are all children of the same, singular God (John 15:13).

> Behold, I send you out as sheep in the midst of wolves; so be wise as serpents and innocent as doves. (Matt. 10:16)

Wolves, sheep, serpents, and doves . . . oh my! That's four shorthand references to animals in one word of admonition and teaching. Sheep are depicted in the Bible as sacrificial animals, free from guile, innocent. Jesus is the Lamb whose blood brings deliverance from death, just as the blood of unblemished lambs brought salvation to the children of the Hebrews in the terrible night of the Passover. Sheep have no natural defenses, depending on safety in numbers and the care of good shepherds. Wolves get no slack in Scripture. They are the bearers of death. To make it in a world of predators, Jesus encourages us to know the ways of the wolf and the serpent, but not to adopt them. It's a challenge to stay innocent as doves when you are trying to offer real life to people who can no longer tell real life from an endless stream of substitutes offered to them by the wolves and serpents of a consumer culture.

You find the word "behold" a lot in these passages. "Behold" means more than merely looking or noticing. It means, "Pay attention, this is important." It means, "Take some time to really see what is before you." Ponder it. Learn from it. Take it to heart. Make it a part of your life's lessons.

You may not think that you can learn anything from an animal. Native peoples the world over would disagree. In their eyes, every creature has a lesson to teach. Eminent scientists and world leaders would also disagree. According to astronomer Carl Sagan:

> The Cetacea hold an important lesson for us. The lesson is not about whales and dolphins, but about ourselves. There is at least moderately convincing evidence that there is another class of intelligent beings on Earth beside ourselves. They have behaved benignly and in many cases affectionately towards us. We have systematically slaughtered them. Little reverence for life is evident in the whaling industry—underscoring a deep human failing . . . In warfare, man against man, it is common for each side to dehumanize the other so that there will be none of the natural misgivings that a human being has at slaughtering another.[4]

Merlin, legendary tutor of King Arthur, knew the value of learning about life from the perspective of animals. I remember with a smile the transformation scenes from *The Sword and the Stone*, Walt Disney's animated version of T. H. White's novel, *The Once and Future King*. Merlin changes young Arthur, nicknamed "Wart," into a fish, a bird, and a squirrel so the future king could experience the world and human beings from their points of view.

Let me invite you to learn from the cetacean nation. Let the whale, the dolphin, and the porpoise give you fresh insight into abundant living as they remind us to breathe, remember our water nature, delight in this life, and give the gift of delight to others. Let our intelligent kin of the sea show us how to cooperate in the unfolding future of the planet, learning the joy in saving lives and making graceful contact with others we too easily, and mistakenly, call "strangers."

> I have been studying the traits and dispositions of the "lower animals" (so called) and contrasting them with the traits and dispositions of man. I find the result humiliating to me.
>
> *Mark Twain,* Letters from the Earth

The Reflecting Pool

The Reflecting Pool

Quote to Ponder

Animals are reliable, many full of love, true in their affections, predictable in their actions, grateful, and loyal. Difficult standards for people to live up to.

Alfred A. Montapert

Questions to Consider

1. In Genesis, do you understand that we are given stewardship over the Creation, to care for it, and honor it as God's? Or is your perspective more one of "dominion"—that all of the rest of the species of life on the planet are here for humans to consume, use, disregard, and kill at will? What difference does it make one way or the other?

2. Some people feel closest to God in nature. Reflect on your experiences of the holy and insights into the nature of God in the midst of the Creation.

3. What are some insights you have gained from paying attention to the behavior of animals?

Actions to Take

1. Reconnect with nature. Go outside.

2. Leave the city, the town square, and the shopping mall. Take a hike.

3. Sit by a stream. Be still and listen for the sounds of life around you.

4. Sail on a lake or some part of the sea. Cut the engine and stop being driven. Be drawn.

5. Go camping. Get up early. Listen.

6. Read folklore from native cultures around the world with an eye toward animal wisdom and totem animals.

Biblical Wisdom

But ask the animals what they think—let them teach you; let the birds tell you what's going on. Put your ear to the earth—learn the basics. Listen—the fish in the ocean will tell you their stories. Isn't it clear that they all know and agree that God is sovereign, that he holds all things in his hand—Every living soul, yes, every breathing creature?

Job 12:7–10 (The Message)

A Breath Prayer

Open my eyes to see you
in all that you have made, O God.

Chapter 2

Cetacean Citings

To the dolphin alone, nature has given that, which the best philosophers seek: Friendship for no advantage. Though it has no need of help from any man, it is a genial friend to all and has helped mankind.
Plutarch

It would appear that we are more willing to consider the possibility of other intelligences on distant planets than we are on our own.
Keith Howell, *Consciousness of Whales*[1]

Can you imagine the presence of another sentient species on planet Earth? One with a longer history in their realm than humans have in theirs (cetaceans have been around three hundred times longer than *Homo sapiens*)[2]—theirs being the interconnected global village of Oceana and ours being the disconnected diaspora of terra firma? Heathcote Williams calls them "a marine intelligentsia" as he reminds us, "From space, the planet is blue. From space, the planet is the territory not of humans, but of the whale."[3]

Are you willing to entertain the notion of differing intelligences? Even within our own species, people are gifted with different kinds of intelligence in differing measure, like emotional, mechanical, artistic, and social capacities that exceed those of others. By many accepted measures, cetaceans possess intelligence that is at least equal in their world to ours in our world—intelligence with all of the prerequisites not only for sentience, but also sapience.

Sentience most simply implies sensitivity, as in perception through the senses. It is the capacity for basic consciousness—the ability to feel or perceive—not necessarily including self-awareness. Sapience indicates knowledge and its application, higher consciousness, and the exercise of will and apperception. Sapience is the cognitive process by which a newly experienced sensation is related to past experiences to formulate an appropriate response. Sapience is self-awareness, choice-making consciousness, and the ability to feel emotions. It involves response-ability, not merely reactive or reflexive behavior.

Cetaceans are definitely sentient beings. They respond to stimuli familiar to us such as touch, sight, hearing, and taste. Their hearing is developed to a much greater degree than their sight, yet they have well-developed eyes and can see (though not in color) both in the water and in the air. Cetaceans even have binocular vision over at least part of their visual field.[4]

Wouldn't it be great if our eyes were largely insensitive to the color of each other's skin, if we were functionally color blind with regard to race, while retaining a sharp and appreciative eye for the rich palate of cultural diversity? Oh, that humans had binocular vision! Then we could see farther down the road. We could see the effects of our decisions as a society on future generations instead of our nearsightedness regarding the use of the earth's resources, the accumulation of personal property, and the inequity of the global distribution of wealth and power.

According to *Encyclopedia Britannica*, "dolphins have been shown to be sensitive to the standard four qualities of taste: sweet, salty, sour, and bitter."[5] Their sense of touch is highly developed and apparently a source of great pleasure to them in interactions with humans and with other cetaceans. Self-stimulation is common in both sexes. Researchers who spend hours in the water with dolphins report high levels of communication through touching skin on skin, and even gentle stroking with teeth on skin.

In 1789, philosopher Jeremy Bentham suggested the principle for establishing the sentience of a creature and, therefore, its entitlement to humane treatment. Comparing certain animals to a human infant, Bentham suggested that our treatment of them cannot be based on our perception of their ability to think and respond, nor their ability to converse with us in our language. In his words, "The question is not, 'Can they reason?' nor 'Can they talk?' but 'Can they suffer?'"[6]

> When I see a dolphin, I know it's just as smart as I am.
>
> *Don Van Vliet*

Cetaceans most definitely feel pain, both the visceral sensation and the emotional variety. They wince and cry out when in physical pain. They mourn the death of family members, evidenced in their vocalizations, their demeanor, and their rituals of carrying the deceased on their beaks for a time before letting them sink to the bottom of the sea or beaching them in an act of burial. They are aware of their pain. They suffer.

Not only do cetaceans meet Bentham's key criteria for sentience, they also appear to reason and act with purpose. And, they do converse—most certainly with each other and to some extent with humans. Just because I cannot understand a bit of Arabic and not much Spanish, it doesn't mean that I can't learn or that those who speak to each other in their native languages do not understand each other. Cetaceans have a highly developed language of whistles, clicks, and glissandos that they use to communicate across vast expanses of ocean in their own kind of World Wide Web.

Intelligence, Genetics, and Soul

Not only do cetaceans perceive the world around them through their senses, they make sense of what they feel, taste, hear, and see. This goes beyond mere sentience, beyond mere sense perception and reflex, to sapience. Sapience may be defined as applied knowledge, or the capacity for judgment and decision-making. Another synonym for sapient is "clever." Sapient, as in *Homo sapiens*. That's right, sapient has the same root as the second word in the biological classification for human beings.

The ancient Greeks recognized that cetaceans breathe air, give birth to live young, produce milk, and have hair—all features of mammals, including us humans. But have cetaceans moved beyond sharing these features with us and other mammals to sentience, or even sapience?

The movie *Bicentennial Man* chronicles the life journey of Andrew, an artificially intelligent android played by Robin Williams, in his quest to be considered human. He is intelligent. He can think, solve problems, and generalize his knowledge to new situations. He is aware of his own existence. But is he sentient? And what is more, is he sapient?

More subtle aspects of sapience include emotional responses like love, fear, empathy, anger, and the development of a sense of humor. In this cinematic treatment of the question "What makes a creature sentient?", the ability to pass gas and swear are also indicators of sapient consciousness. (Other examples of our attempts to answer this question from television, cinema, and literature include Mary Shelley's *Frankenstein*; Isaac Asimov's *The Positronic Man*; the recurring role of Data on the television series and subsequent movies featuring the cast of *Star Trek: The Next Generation*; and Steven Spielberg's *Artificial Intelligence*, originally conceived by Stanley Kubrick). Of course, to really be alive and self-aware, a sapient being must reckon with the question of mortality and deal with the reality of one's own eventual death.

IQ Tests

Let's start with understanding what it means to have the capacity for intelligence necessary for sentience. When we compare the brain of a dolphin with the brain of a human on the cellular level, we discover these three things:

1. The cell count of a dolphin's brain is just as high per cubic millimeter as the cell count of a human brain.
2. The connectivity, or number of cells connected to one another, is the same as that of a human brain.
3. There is the same number of layers in the cerebral cortex of a dolphin as there is in a human brain. In this case it is a good thing to be convoluted!

In other words, [the brain of the dolphin] is as advanced as the human brain on a microscopic structural level.[7]

An essential ingredient in the development of sapient intelligence appears to be social interaction, and social interaction is a catalyst for brain development.

> Anthropologists believe that the development of human intelligence has been critically dependent upon these three factors: brain volume, brain convolutions, and social interactions among individuals. Here we find a class of animals where the three conditions leading to human intelligence may be exceeded, and in some cases greatly exceeded.[8]

Cetaceans are extremely social creatures. At the center of cetacean family life is the mother and child relationship. Studies of coastal dolphins have shown long-term association of dolphins with their mothers. Small nuclear families of mothers and calves and other closely related kin is typically expanded to include additional members who travel, play, and feed together. These expanded groupings are called by many names including school, herd, pod, or gam. These schools can extend over vast areas of the ocean while maintaining contact and communication among its members. Groupings of pods can grow to more than one thousand members! These larger cetacean societies are often maintained over many generations.

Senator Hubert Humphrey quoted Russian dolphinologist Yabalkov at the 1970 U.S. Senate hearings on the Marine Mammal Protection Act:

> Dolphin societies are extraordinarily complex, and up to ten generations coexist at one time. If that were the case with man, Leonardo da Vinci, Faraday, and Einstein would still be alive. . . . Could not the dolphin's brain contain an amount of information in volume to the thousands of tons of books in our libraries?[9]

Cetaceans also appear to enjoy an open and easy intra-class association. Dolphins hang out with whales and seek interactions with humans. We certainly have something to learn here, which would help us overcome our xenophobia—our fear of strangers.

Cetaceans enjoy all kinds of sexual activity and spend a great deal of their time in sensual play. Only a few species appear to engage in sexual activity that goes beyond the instinctive drive for procreation and outside of cyclical breeding seasons. This is another indication of both their highly developed social sense and their sapient intelligence. Lyall Watson and Tom Ritchie state in *Whales of the World*: "If this is so, the sheer quantity as well as the quality, sensitivity, and complexity of sexual behavior in cetaceans puts them very high up in the evolutionary tree."[10]

Gene Genus

The capacity for sapient intelligence in cetaceans is undeniable. But what about genetic makeup? David Busbee, researcher at Texas A&M University, reports in the journal *Cytogenetics and Cell Genetics*, that of the twenty-two chromosomes found in dolphins, thirteen are exactly like human chromosomes, three are strikingly similar, and the remaining six are combinations or rearrangements of human chromosomes.[11] Genetically speaking, we may have more in common with dolphins than with other land mammals. And dolphins may have more in common with humans than humans do with primates.

Cetaceans pass all tests set by anthropologists, geneticists, and bioneurologists for sentient life. In cetaceans we find high intelligence as indicated by brain size, density, and convolutions; social interaction; and evidence of applied memory and learned behavior. We have also discovered that their genetic makeup is similar to the other reportedly highly evolved and sentient species on the planet (that would be us, who supposedly are also able to apply our memory and learn from past experiences). The key word is *supposedly*. Author Douglas Adams writes:

> It is an important and popular fact that things are not always what they seem. For instance, on the planet Earth, man had always assumed that he was more intelligent than dolphins because he had achieved so much—the wheel, New York, wars and so on—whilst all the dolphins had ever done was muck about in the water having a good time. But conversely, the dolphins had always believed that they were far more intelligent than man—for precisely the same reasons.[12]

Soul Searching

But what about having a soul, the test of theologians and philosophers for sapient life? Do porpoises, dolphins, and whales have souls? It depends on what you mean, I guess. To the ancient Greeks, to be alive was to have soul. A being's soul is what gives it being. It is the essence of one's self and self-awareness, the animating force behind thought, emotion, and action.

Plato's understanding of soul included three elements: *logos*, presence of mind and reason; *thymos*, emotion and spiritedness; and *eros*, appetites and desires including simple bodily needs and urges including hunger, love, community, sex, and safety. Although Christians profess belief in the unity of mind, body, and soul—even "the resurrection of the body,"[13]—most live out of a functional belief in an immortal soul that is separated from the body in death, retaining its perception of eternal happiness or torment.

We use the word "soul" to describe the vital core or essential nature of something, someone, or some group. Some use "soul" to name the spiritual or the holy. Other popular connotations of "soul" include the strong demonstrative emotions exhibited by some performers or artists, or a particular kind of ethnic food.

> No good fish goes anywhere without a porpoise.
>
> *Lewis Carroll*

Many people use the words "soul" and "spirit" interchangeably. But, strictly speaking, many languages use multiple words to parse out the different meanings of these two ideas. In Latin, the word *spiritus* refers to both breath and to "soul" in the sense of courage and vigor. "Soul" in Latin is translated *anime,* or the animating force or principle. In Greek, "soul" is translated as *psykhe*, closer to our understanding of consciousness. The Greek word for "spirit" is *pneuma,* or "breath." In Hebrew the word for both "breath" and "spirit" is *ruach*. The Hebrew word *nephesh* describes the breath of life that God breathed into the first man and woman, *adam* and *adama*.

In Jewish teaching, *neshamah* is the uniquely human soul that gives humans spiritual awareness and the possibility of relationship with God. This distinction underscores the belief that man is essentially different from the rest of the species that share the planet, including other mammals. As Psalm 8:5 declares, God has made us a "little less than the angels."

Some take this as a sacred trust to act as stewards of God's Creation and all of its creatures. Others take it as a license to disregard our interdependent ecosystem and simply use God's

Creation and all of its lesser creatures for the comfort, benefit, and indulgence of man. But I wonder, do we not all share the breath of God? Is not life, life? Is there not one enlivening spirit, the Holy Spirit? Does not all that lives live because of the same breath of life?

You may think it's going too far to say that other sentient mammals have a soul (although there are a lot of us who have offered comfort to a child after the death of a pet with the assurance that Rover is indeed in doggie heaven). But, we can certainly agree that cetaceans are spirited creatures! According to Scott Taylor:

> If a "soul" can be defined as an aspect of inner divinity, a sparkling inner Self who knows the warmth of love and joy that comes from being alive, who reflects this awareness in its harmonious life by choosing to do so, then there are souls in the sea.[14]

Creative Creatures

Other markers for sapient life include the ability to create art and the use of tools. According to Don White, cofounder of Project Delphis (a dolphin research and advocacy group formed in 1985 with Dexter Cate), dolphins create beautiful symmetrical rings and spirals, some up to twenty feet long, out of water and air! White's description is fascinating:

> The young dolphin gives a quick flip of her head, and an undulating silver ring appears—as if by magic—in front of her. The ring is a solid, toroidal [donut-shaped] bubble two feet across—and yet it does not rise to the surface! It stands erect in the water like the rim of a magic mirror, or the doorway to an unseen dimension. For long seconds the dolphin regards its creation, from varying aspects and angles, with its vision and sonar. Seemingly making a judgment, the dolphin then quickly pulls a small silver donut from the larger structure, which collapses into small bubbles. She then "pushes" the donut,

> which stays just inches ahead of her rostrum, perhaps 20 feet over a period of up to 10 seconds. Then, stopping again, she regards the twisting ring for a last time and bites it—causing it to collapse into a thousand tiny bubbles which head—as they should—for the water's surface [but only after a slight, surreal delay in their upward movement]. After a few moments of reflection, she creates another.[15]

This isn't fantasy, it's real. And it isn't magic, just marvelous. This is no mere instinctive reflex. It is rare behavior that illustrates the dolphins' ability to control the elements of their environment, apparently by choice, for their own enjoyment and the enjoyment of others. It is a learned skill passed from one dolphin to another, as a teacher to a student. These temporary artifacts can't be recreated by humans in real time with real water, but only with the aid of computer-generated animation! If art can be defined as self-expression in the creation of artifacts that are then experienced, reflected upon, reacted to, and appreciated for their own sake, then these magnificent rings are definitely art.

> Listen to the voice of a dolphin, and you shall learn the secret to mankind's survival: PEACE!
>
> *Mallory Watson*

The silver rings and spirals are air-core vortex rings generated with a whip of a dolphin's dorsal fin or its head and a well-timed infusion of air from their blowhole. A phenomenon called the Bernoulli effect accounts for the ability of the rings and spirals to keep their shape until they are dispersed by the snap of the dolphin's mouth. The mysterious delay in the upward movement of the tiny air bubbles created in the "popping" of the air-core vortex (the silver donut) is caused by the residual imbalance of pressure around the bubbles created by the dolphin artist in the first place.[16]

Although the laws of physics can explain these silver rings of water and air, their spontaneous creation by sapient cetaceans, not as an instinctive reflex but as a learned skill, is awe-inspiring and instructive. Researcher Don White puts it this way:

> As evidence mounts for "self awareness" and other "intelligent" qualities in dolphins, I think that it must cause us again to ask the question: what are these creatures, that they spin silver lariats for the sheer joy of creation? And what sort of creatures are we, if we cannot appreciate and protect them?[17]

Tool Time

In our human-centric study of the planet and its civilizations, we have made the presence of technology, the use of tools as extensions of ourselves, a hallmark of intelligence and sentience. Susan Milius, a reporter for *Science News*, offers compelling evidence that dolphin mothers actually teach their young to protect their sensitive beaks with sponges as they search for food in reefs and along the ocean floor. According to Milius, this is a learned behavior, probably passed from mother to child, rather than the instinctive response resulting from inheriting some sponge-wearing gene. After extensive study of the sponge-using dolphins in their only natural habitat, Shark Bay in Australia, genetic researchers from the University of Zurich in Switzerland concluded that dolphins' sponge use is a case of cultural transmission—the passing along of a learned behavior.[18]

The use of tools as an extension of our bodies to perform at a higher rate of efficiency, safety, or skill is a sign of forethought, intention, and invention. We use pencils and pens to write; racquets, bats, and paddles for sport; and screwdrivers, hammers, and forceps for greater dexterity and the manipulation of our environment. It is true that we do not observe cetaceans erecting buildings or manufacturing countless gadgets that entertain or serve some sense of fad or style.

Maybe, as Heathcote Williams suggests, they have invested all of their considerable brain power and length of life on the planet in a different sort of intelligence with a different set of values and goals. Maybe even a lifestyle from which we can learn. To quote Williams:

> Without hands, for making external tools,
> Without depending on tons upon tons of artifacts, destined
> as litter
> The whale's sophistication has become internal:
> Its skills are all perceptive,
> Social, sensual, jokey,
> Non-manipulative.[19]

What's in a Name?

The family of species in the order Cetacea includes whales, dolphins, and porpoises. The order gets its name from the Latin word for whale, *cetus*. All of the species in the order Cetacea technically are whales, including porpoises and dolphins. What we commonly call whales are classified as Mysticeti, or baleen whales, one of the two suborders of Cetacea. Dolphins and porpoises are classified in the suborder Odontoceti, or toothed whales. There are more than forty species of whales, thirty-six species of dolphins, and six species of porpoise in the cetacean nation.

The porpoise gets its name from the combination of two Latin words, *porcus* meaning "pig," and *piscis* meaning "fish"—or, pigfish. Of course, it is neither a fish nor a pig. It's a mammal. But the etymology of "porpoise" reminds me of a quote from Winston Churchill: "I like pigs. Dogs look up to us. Cats look down on us. Pigs treat us as equals."[20]

The word "porpoise" entered our English dictionary as an intransitive verb in 1909, meaning "to leap or plunge like a porpoise." Its second meaning, "to rise and fall repeatedly,"[21] is an invitation to embrace the hopeful possibility of resilience and remember the sheer persistence of life!

Most people use the terms porpoise and dolphin interchangeably. However, there are differences. Porpoises are not as sociable with humans and they tend to travel in fewer numbers than dolphins. They have small, rounded heads and blunt snouts instead of the bulbous "melons" and longer beaks of dolphins, and they tend to be smaller and darker in color than dolphins.

> To think the way we do [the whale] would need to use about one-sixth of his total brain.
>
> *J. Lilly, from* Man and Dolphin

The Latin family name for dolphins within the order Cetacea is *delphinidae*, derived from the Greek word for dolphin, *delphis*, which is closely related to the word for "womb." The womb, of course, is understood as the source of new life, the incubator of creation, and a wellspring of wisdom. The famous oracle at Delphi gets its name from the god Apollo (god of music, prophecy, poetry, medicine, and competitive archery), whose cult title was Apollo Delphinios. There are several versions of the story of Apollo's acquisition of the ancient oracle, once the province of earth-mother Gaia, guarded by the dreaded serpent, Python. In one, Apollo came to Delphi (known before him as Pythia) disguised as a dolphin, bringing his priests from Corinth. In another, a dolphin carries Apollo to Pythia.[22]

The God of Abraham, Isaac, and Jacob endowed all creatures with life and encourages us through the words of holy Scripture to learn from them. So, why not cetacean wisdom? Why not look for insight from dolphins, whales, and porpoises?

After hearing Peter Shenstone, renowned carrier of the modern oral tradition of dolphin lore, an Australian Aboriginal sage named Burnam Burnam remarked:

> Well done Peter . . . you have done what no one else has ever done. You have found the totem animal for the White Fella! The dolphin is the only creature that every white person will take notice of. People of color have always had totems, special animals that teach us. Now the White Fellas have one. Good on ya, mate![23]

Can we imagine another way to measure intelligence and applied wisdom that may be different from but equal to humankind's? Is progress and accomplishment to be measured by physical monuments, sociological constructs, and technology alone? If intelligent civilization is measured by the ability to thrive without killing our own species, managing a sustainable and globally distributed food supply, taking empathetic care of others in a highly functioning community, creating art and music, and living life with a profound sense of joy, then I can imagine it. We can observe it, learn from it, and participate in it.

Williams puts it this way:

> The god Apollo saw the smallest whale, the dolphin,
> As the embodiment of peaceful virtue, undisguised joy,
> And as a guide to another world.
> He sometimes exchanged his god-like status
> To assume dolphin form;
> And formed the oracle at Delphi,
> Named in the dolphin's honor.
> There, the god hoped,
> Man might be guided by a sense of other-worldliness.[24]

This "other world" is not some ethereal fantasy world, but a parallel world coexisting on this planet. It is Waterworld, Oceana, home of the Cetacean Nation. We are literally surrounded. Surrounded, as it were, by a great cloud of witnesses.

The Reflecting Pool

Quotes to Ponder

If having a soul means being able to feel love and loyalty and gratitude, then animals are better off than a lot of humans.

James Herriot

And when the day comes that we can communicate intelligently with dolphins, they may introduce us to the concept of survival without aggression, and the true joy of living, which presently eludes us. In that circumstance what they have to teach us would be infinitely more valuable than anything we could offer them in exchange.[25]

Horace Dobbs, Follow a Wild Dolphin

Questions to Consider

1. Why do you suppose we continue to work toward developing "artificial intelligence" that mimics and perhaps will surpass our own?

2. What can we learn about ourselves by studying and observing cetaceans?

3. What makes a human, human?

Actions to Take

1. Read Psalm 8 and ponder its implications.

2. Find out all you can about cetaceans through books, articles, television shows, the Internet, conversations with experts, and your own observations.

3. Watch *Bicentennial Man* or *Artificial Intelligence* with a group of friends and discuss it afterward.

4. Play a game of "What If" by asking, "What if there were other, different intelligences in the universe, or on our own planet, equal to or surpassing that of the human species?"

Biblical Wisdom

And God saw everything that he had made, and behold, it was very good.

Genesis 1:31

A Breath Prayer

Gracious God, help me do all I can
to become more fully human.

Chapter 3

Hydrate

Unless one is born of water and the Spirit,
he cannot enter the kingdom of God.
John 3:5

Let justice roll down like waters,
and righteousness like an ever-flowing stream.
Amos 5:24

The dolphin, the fish that is not a fish, is inseparable from the imagery, persona, and theology of Jesus Christ and the church. Jesus is called *Ichthus.* This Greek word for "fish" is also the ancient abbreviation for the words "Jesus Christ, Son of God, Savior."

In the early days of Christianity, especially during the persecution of Nero, the sign of the fish was a signal that one Christian was approaching another. One would draw a half-circle arcing outward in front of him in the dirt. The second would draw an identical half circle that began at one end of the first arc and overlapping it slightly at the other end to form the shape of a tail fin. The sign of the fish was the dominant symbol of Christ and the church until the time of Constantine (in the third century), when it was superceded by the sign of the cross.

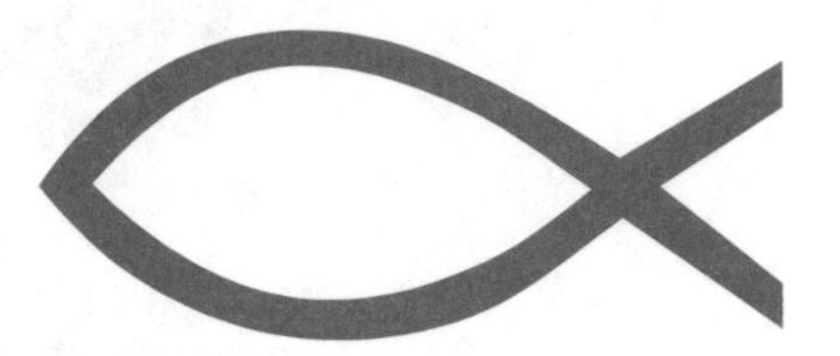

Jesus ushered in the age of the servant community, one in which people supported one another, sharing all things in common, raising the rule of love above the rule of law. There is no law greater, and all laws would be unnecessary if we would love God and love our neighbor in all things. Some refer to this as the Aquarian Age, the age of water, heralded by one who pours out the waters of life. Jesus, the Dauphin[1] of God, promises living water to anyone who seeks it (John 4:7–15).

Porpoises and their relatives invite us back to the water. Remember, we were water creatures for the first nine months of our lives. Cetaceans live most of their lives under water, but they have to come up for life-sustaining air. We spend most of our lives out of the water, but we would not have life without the water within us. Jesus teaches that we must be born of water and the Spirit. The porpoise invites us to revisit the water, elixir of life, to experience a renewal of our spirits.

Water, Water Everywhere

From beginning to end, the Bible is saturated with water images. True to its nature, water is experienced in a variety of settings with a variety of effects. It is assigned sometimes-contradictory meanings and brings both life and death, yet it is all water.

One Sunday morning my wife, Gloria, and I joined our friends Ed and Kathleen Kilbourne in leading a service of baptism and renewal with the Methodist congregation of the island village of

Hopetown, in the Abacos of northern Bahamas. The entrance to the beautiful chapel faces in the direction of the harbor, just inside the Abaco Sea. The windows behind the chancel, where the altar table rests, face out into the Atlantic Ocean.

In an effort to add layers of meaning to the service, Kathleen asked our young friend, Eli (who was about eight years old at the time), to collect some water from the harbor and some water from the ocean to mix together in the baptismal font. The exasperated Eli responded, "You know, it's all the same water." To which the now-enlightened Kathleen replied, "Yes, but the water in the harbor helps us remember water as a source of safety and the home of our community. The water from the ocean helps us remember the wideness of God's love, and the thrill and danger all around us. It's all the same water, yet it has different meanings in different places."

> Anyone who can solve the problems of water will be worthy of two Nobel prizes—one for peace and one for science.
>
> *John F. Kennedy*

Creation begins with water, the main ingredient of the primordial soup. "In the beginning God created the heavens and the earth. The earth was without form and void, and darkness was upon the face of the deep; and the Spirit of God was moving over the face of the waters" (Gen. 1:1–2). And the waters brought forth life, both in the seas and upon the land. The Tigris and Euphrates rivers, together with the lesser-known Pishon and Gihon rivers, flow from the Garden of Eden to nourish all living things. Today, all life begins and finds its sustenance in water.

Life is threatened when the water dries up and fails to fall. No doubt, you have experienced both physical drought and spiritual drought. Yet, God promises spiritual resources that will not only sustain you but those around you, when he says in Isaiah 58:11, "and you shall be like a watered garden, like a spring of water, whose waters do not fail." Hear this prophecy from Revelation 7:17, " . . . and he will guide them to springs of living water; and God will wipe away every tear [more water] from their eyes."

> Water is life's matter and matrix, mother and medium. There is no life without water.
>
> *Albert Szent-Gyorgyi,*
> *Hungarian biochemist and*
> *Nobel Prize winner for medicine*

The gift of water becomes a deadly curse as the Great Flood devastates the earth as recounted in Genesis, chapters 6 and 7. The waters of the Red Sea become both a gateway to freedom and new life for the Hebrew people and a deadly grave to those who would enslave them in the same sweeping episode in Exodus 14:21–29. Today, we know the devastating force of tsunami and flood, of hurricane and typhoon.

Sometimes bodies of water can call forth feelings of melancholy and longing for loved ones and what is familiar, like the Israelites felt in exile by the waters of Babylon (Ps. 137:1). And yet, we know that it is water that connects us all, and in a global sense, it's all the same water. God leads us beside still waters (Ps. 23:2) and is with us in the stormy seas (Isa. 43:2).

There is the river that makes glad the heart of God, described in Psalm 46:4. This river has many streams, each with its unique contribution to the fullness of God's satisfaction: streams of justice, mercy, and righteousness. The prophet Amos proclaims the word of the Lord, "Let justice roll down like waters, and righteousness like an ever-flowing stream" (Amos 5:24). Access to clean water is already a global justice and health issue. Futurist/historian Leonard Sweet predicts that futurepresent struggles will be for water, not oil.[2] To offer a cold cup of water in Jesus' name (Matt. 10:42) takes on new dimensions of justice and mercy in this global context. Offering water to one who is thirsty or who does not have access to clean water is the same as offering it to Jesus (Matt. 25:35–36). The Bible begins with life springing forth from the waters and ends with the free offering of the water of life to all who are thirsty in Revelation 22:17.

United Methodist Bishop Will Willimon (back when he was a young professor at Duke Divinity School and my faculty adviser) suggests that baptism conveys all of the things that we associate with water: birth, life, death, rebirth, cleansing, and joy. The reenactment of dying to self and sin and being resurrected to new life in Christ is most literally experienced in baptism by immersion, when one is "buried" beneath the water and raised up a new person, cleansed of sin and walking in new life.

Jesus speaks of the waters of birth and rebirth in John 3:5, saying you must be born of water and the spirit. Jesus submits to baptism at the hands of another, that he may also be both born and reborn in water, rising to hear these words, "This is my beloved son, with whom I am well pleased" (Matt. 3:13–17).

> You'd be surprised how many things
> Are dry and useless till one brings
> The magic liquid known to all;
> You use it when you heed the call—
> Just add water.
>
> *David J. Ford,* "Just Add Water"

H_2O and the Marriage Mystery

Two hydrogen atoms join together with an oxygen atom and the three become a new creation. Genesis 2:24 is often read in traditional wedding ceremonies as we affirm that "the two become one flesh." But, which one do they become? In a new twist on a well-known formula, which one does the combination of Ha! (one partner)+Ha! (one partner)+Oh! (the Holy Spirit) equal? Answer: neither of the two! Just like when hydrogen atoms and an oxygen atom join together to make something that is neither hydrogen nor oxygen, the two become a new creation with Christ as the catalyst in their midst. They become living water for each other, and others in their web of relationships, in Christ—an embodiment of Christlife together.

Water Nature

We are mostly made up of water. The ratio of water to other substances in our bodies is about the same as the ratio of water to land on the earth, about 70 percent to 30 percent. It is an essential ingredient in the formation of life as we know it.[3] If judged by volume, carbon-based life is actually water-based life.

We are intrigued by the discovery of water on Mars as a sign of the possibility of past, and even present, life. Arthur C. Clarke speculates about the possibilities of sentient plant life on one of Jupiter's moons, the icy Io.[4]

Since we are made in God's image, God must have, at some level, a "water" nature—a fluid, dynamic, life-birthing, earth-cleansing, powerful, still, and chaordic (order in apparent chaos) nature. "Go with God" could, in some cases, be translated as, "Go with the flow!" Or more precisely, "Get in the flow."

Water is a great object lesson for illustrating the trinitarian realities of God. Water is essentially the same but existentially different (like matter and energy, $E=mc^2$), manifesting itself differently as a liquid, a solid, and a gas depending on the circumstances. Why should we have such a hard time with changing our methods if our goals and values stay the same? I can be essentially the same person at all times while adapting to the shifting realities around me. I am a father, son, husband, brother, and a citizen of at least three "kin-doms" (the Kingdom of Heaven, Earth, and the United States). I am a pastor, a musician, a disciple of Jesus, and a friend. I am most healthy and most present in all of these roles if I remain essentially me as I inhabit each of them.

Water Logic and Rock Logic

Educator and creativity coach Edward de Bono identifies Aristotle as the progenitor of inclusion/exclusion logic, or "box logic," that is the basis of so much of our ingrained decision-making processes.[5] In this kind of "judgment thinking," everything has to fit into an "either/or" category in a sorting-out process based on deciding what goes in a designated box and what does not. (This is the famous box we are being encouraged to think outside of.)

There is no other choice in this model. Everything on the proverbial table either goes in the box or it does not. When this kind of logic is used in decision-making, someone is always right and someone else is then wrong. There is a right answer and a wrong answer. This tabletop box logic serves us well in many cases; we can

> Nothing on earth is so weak and yielding as water, but for breaking down the firm and strong it has no equal.
>
> *Lao Tzu*

determine, evaluate, and then predict with a high degree of certainty and confidence across a wide variety of standard situations. Some examples: "A chemist can predict the behavior of a chemical. The judge can dispense justice."[6] We want our doctor to decide what ails us and to prescribe the correct therapy.

According to de Bono, "The weakness of the judgment system is that it was never designed for change."[7] Organizations see no need for change if they have reached a stable state. But if conditions change and standard isn't standard anymore, the old judgment system will not work. "Judgment is basically a backward-looking [pastpresent] system. This is enough for most of our thinking and behavior, but we also need [a logic system that provides for] 'forward-looking' [futurepresent] design and innovation."[8]

> If there is magic on this planet, it is contained in water.
>
> *Loren Eisley*

The old judgment system of either/or thinking can be called rock logic.

> A rock is solid, permanent, and hard. This suggests the absolutes of traditional thinking (solid as a rock). Water is just as real as a rock but it is not solid or hard. The permanence of water is not defined by its shape.[9]

What does this have to say about your normal response to change? Are you highly adaptable? Can you, like the porpoise, go with the flow? Do you live in the reality of a fluid environment or are you in a state of denial, clinging to your boxes and "right" answers?

> A rock has hard edges and a definite shape. This suggests the definite categories of traditional [either/or] thinking. Water [also] has a boundary or an edge . . . but this boundary will vary

> according to the terrain. Water will fill a bowl or a lake. It adapts to the terrain or landscape [while simultaneously shaping that terrain or landscape]. Water logic is determined by conditions and circumstances.[10]

The terrain changes, the water flows into it, but it remains essentially water.

Rock logic is based on what *is*. Water logic is based on flow, so we use the word *to*. A rock *is*. Water *flows to*. Rock logic asks what is right or wrong. What is or isn't?

But we don't live in an either/or world anymore. Remember the world-changing question, "Does light behave as a particle or a wave?" Answer: "Yes."

Water logic asks *what fits* and *where does it flow*, or lead us. *Fit* and *flow* are our simultaneous questions. If we define truth as "a particular constellation of circumstances with a particular outcome" we see "both concepts of fit (constellation of circumstances) and flow (outcomes)."[11]

If all this seems too pragmatic and "relative," remember that water has well-defined behavior—it follows rules. For example, water will not flow uphill and it will follow the gradients in the landscape.[12] And, it always remains water.

I'm not saying that water logic is better or more right than rock logic. I'm not debating the worth of one over the other. I'm describing two helpful ways of thinking and deciding. It's not either/or. It's both/and.

Come to the Water

Jenny Cannon, a member of my doctoral project community, offered this meditation on water:

> Someone asked the question today in class, "What is your favorite way of encountering water?" I thought it was intriguing, and the varying responses by people pointed to the ways it spoke to all of us.

> Water links us to our neighbor in a way more profound and complex than any other.
>
> *John Thorson*

> I don't want to oversimplify, but the concept of water seems to be a pretty powerful image for God. Water itself seems very powerful, whether through actual dangerous physical force or the emotional impact of a stream, or even the spiritual symbolism of baptism. Water forms around us, shapes around our bodies, and when we share water space with someone, it makes room for us all. Water is life sustaining and yet also has the potential to completely engulf us, even fatally.
>
> I don't [know] all the ramifications of this metaphor, and I realize it breaks down at several points, as do all of them. But in thinking about water and the ways in which we encounter water, it seemed an interesting way to think about the ways we encounter God, both individually and in community.

This way of thinking based on the metaphor and characteristics of water resonates with many of the sensibilities and qualities of the porpoise-given life: Fluidity. Adaptability. Discerning what we need to hold on to and what we need to release. Essential stability in existential change.

The scene of a dolphin riding the bow wave of a boat, or a pod of dolphins playing leapfrog in the water, is an inviting scene. With them, I invite you to walk, stand, or sit by the waters of lakes, rivers, and seas. I invite you to touch the water and remember your essential nature. And if you've been baptized, remember your baptism. I invite you to participate in the politics of water and to do whatever you can to make sure people don't die from the water they drink. I invite you to revisit the joy of jumping into, gliding through, and floating on water. There is a world of difference between contemplating water and jumping into it!

"The Water"

by Chris B. Hughes
from the CD *Hourglass*[13]

Let me see the water
Flowing free flowing free
Let me see the water flowing free

Let me hear the water
Calling me, like the sea
Let me hear the water calling me

Baptized in the name
Of the Father, Son, and the Holy Ghost

Let me feel the water
Rain on down, Rain on down
Let me feel the water rain on down

Raise me from the water
To higher ground, higher ground
Raise me from the water to higher ground,

Baptized in the name
Of the Father, Son, and the Holy Ghost
Let me hear the water, calling to me

The Reflecting Pool

The Reflecting Pool

Quotes to Ponder

When you drink the water, remember the spring.

Chinese Proverb

A lake is the landscape's most beautiful and expressive feature. It is earth's eye: looking into which the beholder measures the depth of his own nature.

Henry David Thoreau, Walden, *1854*

Questions to Consider

1. Why do we spend billions of dollars on bottled water when most of it is the same as tap water? Did you know it costs about $150 to install a clean-water well in most two-thirds world communities? (I use the term "two-thirds world" because actually more than two-thirds of the world's societies are considered developing with their populations falling below international poverty levels.)

2. What are the hard edges of your life that could be softened by water logic?

3. What do you learn about your own nature as you look into the waters of a lake, a stream, or an ocean?

Actions to Take

1. Sit beside the water and let your mind wander and your spirit rest.

2. Seek out or instigate a baptism renewal service or a ritual cleansing service. All you need is some water, some friends, and a willing spirit. Liturgies are available in many hymnals and prayer books (and even online; search under "baptism renewal service"). A designated holy person is not a requirement.

3. Go swimming, preferably skinnydipping. If you don't know how, learn. Face your fear of water. Find a trusted friend or professional to help you literally get back in the water.

4. Investigate the global water crisis, the bottled water scam, and the organizations that provide cheap safe-water systems to two-thirds world communities. Then do something.

5. Go snorkeling, preferably around a coral reef. Swim with dolphins!

Biblical Wisdom

As a hart longs for flowing streams,
so longs my soul for thee, O God.
My soul thirsts for God,
for the living God.

Psalm 42:1

A Breath Prayer

Let my life seek its Source today and all days.

Chapter 4

Listen

To draw near to listen is better than to offer the sacrifice of fools . . .
Ecclesiastes 5:1

He who has ears to hear, let him hear.
Luke 14:35

There is a pivotal scene in the movie *Michael* when the main character, the very unangelic angel of the movie's title, tells a woman all about herself. When she asks him how he knows so much, Michael replies, "I pay attention."[1] Cetaceans pay attention by a continuous process of listening. Although they can see well enough and exhibit active senses of taste and touch, in the case of dolphins, porpoises, and whales, the ears have it.

Talk about being wired for sound! The human brain averages about 2.6 pounds. The average weight of the brain of a bottle-nosed dolphin is four pounds. The record weight of cetacean grey matter is more than twenty pounds! Given the all-encompassing function of sound intelligence

necessary for cetaceans to survive in their world, they may well use their considerable brain capacity for processing acoustic information.[2] "Their whole body: every bone, every membrane, every hollow, [is] part of an enormous ear, twenty times as sensitive as man's."[3]

All cetaceans produce sound. Some of it is produced at frequencies we can hear, and some of it well above and below the frequencies that can be picked up by human ears. Cetaceans make a wide variety of sounds—clicks, pops, whistles, moans, and whines—without benefit of vocal cords (which are restricted mostly to primates). With the aid of nasal passages, laryngeal pouches, air sacs, and fatty tissue in their noses and near their blowholes, sea mammals make noises for at least two reasons. One appears to be communication with other cetaceans. The other is for "seeing."

All of the sounds cetaceans make are meant more as expression than communication, defined as at least a two-way exchange of meaning. Some of the sounds appear to be expressions of emotion, like fear, joy, and affection. And, although communication does not necessarily imply language, cetaceans definitely engage in plenty of what might be called "call and response" conversation. They send out mating calls, distress signals, and "come-and-get-it" mealtime messages. Their low-frequency sounds are powerful enough that some whales can communicate across entire ocean basins, creating their own version of a wireless World Wide Web.

Cetacean GPS

The technical term for using sound to see is "echolocation." Dolphins and porpoises generate certain sounds, sometimes in stereo, from two cerebral cavities in their melon-shaped heads. They listen to the echoes to help them both orient themselves and navigate in the water, distinguishing the objects and creatures around them. It's like natural sonar. These toothed whales (remember, porpoises and dolphins are in the whale family) use extremely high frequencies to refine spatial resolution from their echoes, effectively creating a virtual 3-D picture of their surroundings. They swim less by sight, more by sound.

They catch squid and hunt fish in total darkness, navigate narrow channels and inlets at full speed, and find their way back to sea without bumping into the banks or other obstacles, including each other! The finely tuned creatures can detect and identify objects as small as .015 inches in diameter, the size of a shrimp's feelers! Heathcote Williams poetically describes it this way:

> The whale moves in a sea of sound:
> Shrimps snap, plankton seethes,
> Fish croak, gulp, drum their air bladders,
> And are scrutinized by echo-location,
> A light massage of sound
> Touching the skin.[4]

Orient Express

The first step in orienteering is to get a fix on your present location. Once you know where you are, you have a point of reference for where you want to go. And, others may be better able to reach you if they have a fix on your location. Of course, this may be a physical location or a philosophical, ideological, or metaphysical location. We are trying to locate each other when we say things like, "I just don't know where you're coming from," or, "From where I stand, I see things this way . . . "

A compass is a handy tool for determining true north, or at least as true as possible given the variables involved in relying on the earth's magnetic field. Birds and fish use magnetoreception in the course of following their migration routes. Some researchers believe that a misfire in magnetic detection may be the culprit in certain cases of mass stranding and beachings of large numbers of cetaceans.

You could say that this stranding is caused by an attraction to some "false north." Priest and mystic Matthew Fox, in his winsome book *WHEE! We, wee, All the Way Home,*[5] has developed a life compass. Its true north points toward life and all things that enhance life, protect life, celebrate life, and create life. This true north orientation leads to wholeness and the abundant life Jesus embodies and offers to all people, even all creation. Its direct opposite point is death and all that diminishes life, devalues life, and destroys life—your life, the lives of others, even the life of the creation.

> Life has taught us that love does not consist in gazing at each other, but in looking outward together in the same direction.
>
> *Antoine de Saint-Exupéry,*
> *author of* The Little Prince

To the northwest, in the general direction of the true north of life, are what Fox calls "beauties." These beauties include the natural ecstasies of nature, sex, creativity, friendship, exercise, play, and re-creation. The beauties of life include sensuality—the enjoyment of food, tactile sensations, the wonders of sight and sound. And let's not forget thinking and imagination. They also include servanthood, altruism, and passionate appreciative engagement with diverse peoples and cultures. Experiences of ecstasy also come with practices of voluntary deprivation like fasting, seasons of abstinence from favorite activities or comforts, and from rituals like chants, prayers, and meditation.

To the northeast of life, Fox's compass points toward justice—the work of ensuring that all people, not just the privileged, relatively wealthy, or powerful have a shot at life, and life more abundant.

To the southwest and southeast of death, the insidious south, certain death, are lands inhabited by life-stealing dragons—dragons of bloated egos, institutions that have lost their birthing passion and that perpetuate a smug paternalism, and vicarious, voyeuristic living filled with excuses and second-hand experiences. In these regions of death lurk the dragons of unnatural ecstasies—alcohol abuse, external validation, and consumerism.

These cunning dragons perpetuate the dichotomies of male and female, first world and third world, black and white, Hispanic and Asian.[6] These are the very socioeconomic and cultural dichotomies the Gospel of Christ comes to dissolve (Gal. 3:28). These false separations and categories undermine movement toward the true north, toward wholeness, cooperation, and inter-dependence, toward true abundant life.

How would you describe true north on your life's compass? What values or goals keep you moving in a life-affirming direction? What draws you and attracts you? Are you drawn to natural ecstasies or to unnatural ones? Do you care about the greater good of your community or is life all about "me and mine?" Do you ever get righteously angry because someone you don't even know is unfairly treated or because some public policy or business-as-usual practice creates a growing number of working poor in our midst? Getting off course, drifting toward the insidious south on the compass of life, can lead to the stranding of yourself and of others who may be following you or traveling with you.

Remember the first step in orienteering—figuring out your present location. Where are you now, emotionally, spiritually, physically, and relationally? Are you self-aware enough to know

> Failure is only a temporary change in direction to set you straight for your next success.
>
> *Denis Waitley*

and name the state you are in? Is it where you want to be? If not, where do you want to be? Do you see a path forward? Can you chart a course from where you are to where you want to be? Remember, cetaceans rarely go it alone. They literally use others in their pod and obstacles in their path as sounding boards as they navigate through their world.

The echoes of your past may be creating so much static and false feedback that you are caught in a loop of guilt or failure. You may not be able to see around the obstacles ahead. Use the sounding boards around you. There are friends, family members, counselors, pastors, and mentors who are willing to help you get re-oriented, test your version of reality, and help you move ahead with renewed confidence and joy. Here's singer and folk theologian Ed Kilbourne's version of an old song that celebrates the confidence of reliable echolocation:

I know where I'm going
And I know who's going with me
I know why there's music in the quiet summer morning
I've found a wealth of gold
And silver have I plenty
I've found a light to guide me
When my way gets dark and stormy

Where are you going
And who will walk beside you
When the night is gloomy
Where is the light to guide you
Where is your gold
And silver brightly shining
Who writes the music in the quiet summer morning

I'm going where He goes
And He'll be there to guide me
The love for which he died

Is all I need beside me
And he's my gold and my silver brightly shining
He writes the music in the quiet summer morning

I know where I'm going
Where are you going?[7]

I'm Looking Through You

You've heard the expression, "I can see right through you." Well, cetaceans really can! Using a constant stream of sound, they can "see" through skin, scales, muscle, and fat to bone, with the ability to distinguish friend from foe and food from family. With their penetrating gift of echolocation, cetaceans really can see through other creatures and, in turn, are literally transparent to other cetaceans.

The process of echolocation involves what Williams describes as:

> intense salvoes of bouncing clicks, a thousand a second, with which a hair as thin as half a millimeter can be detected [and identified]; penetrating probes, with which they can scan the contents of a colleague's stomach, follow the flow of their blood, take the full measure of an approaching brain.[8]

This sounds like techniques used in lie-detection and stress tests, as we depend on sophisticated equipment external to ourselves to detect small changes in the retina, blood pressure, respiration, and perspiration to signal emotional states and judge a person's truthfulness.

Cetaceans signal changes in their emotional state through intestinal peristalsis (the rippling motion of muscles in the digestive tract) the way we do through facial expressions. The difference is, we can control our facial muscles and disguise our true feelings. Dolphins can't hide their true feelings. Their guts give them away. They can't be disingenuous. They can't lie. It would take some getting used to, but imagine the dynamics of a truly genuine and transparent community!

Be Still and Know

For many people, praying involves a lot of talking (or at least running) through a list of people and concerns in their mind. Prayer lists are helpful, but they are not the first step in developing an effective prayer life. A more helpful first step is learning to be still. The prophet Elijah, fleeing for his life to Mt. Horeb, heard God speaking in a still, small voice. This is not to say that God cannot be heard in the raging winds and the blazing fires of life. But there is a clear call to be still and know God (Ps. 46:10). On their *Hell Freezes Over* album, The Eagles offer their invitation to "Learn to Be Still."

> Though the world is torn and shaken
> Even if your heart is breakin'
> It's waiting for you to awaken
> And someday you will—
> Learn to be still [9]

Know what you are listening for. Mechanics, musicians, and sheep are all great listeners. Car mechanics always ask me what my engine sounds like in their effort to diagnose the problem. They have specific sounds (chugs, pings, clicks) that they ask me to imitate to them over the phone.

Musicians learn to "talk drum," mimicking the ka-chings, kuh-tishes and boom-picka-taos so they can communicate the sounds and rhythms they are looking for to their percussionists. (In fact, Japanese taiko drumming has no written music—only special sounds that musicians use to teach the compositions. Amazing that an ancient percussive tradition has been passed down completely orally!)

And as for sheep . . . remember, they know the voice of their master and can distinguish it from all others. Shortwave radio enthusiasts learn to pick out the voices of their friends around the world through the almost constant static of their receivers. It all depends on knowing what you are listening for.

A Spiritual V8 Moment

A young boy named Samuel was born to an aging couple. They had promised to give their child back to God if God would give them a child. True to their word, the couple brought their son to the temple to be raised by a priest named Eli.

The child grew into a young man, approaching the age of his bar mitzvah. One night, he was asleep in the temple where the Ark of the Covenant was kept, and he was awakened by a voice calling his name, "Samuel, Samuel!"

Young Samuel thought it was his master, Eli, calling him. He ran to Eli's room and presented himself saying, "Here I am, for you called me."

Now, Eli had not called the boy, so he sent Samuel back to bed. This happened again, and a third time, whereupon Eli slapped his forehead in a spiritual "I-could've-had-a-V8" moment, realizing that it was God calling out to Samuel. He told Samuel, "Go, lie down; and if you hear the voice again say, 'Speak, Lord, for your servant hears'" (1 Sam. 3:1–9, paraphrased).

There are many voices calling out to you, and there are too many people telling you when it is God. Yet, God does speak: in still, small voices; in the cries of the needy; in the counsel of faithful friends; and in the convergence of circumstances that call out for some response from you.

> Our thoughts create our reality—where we put our focus is the direction we tend to go.
>
> *Peter McWilliams*

Would you know the voice of God if you heard it? When do you know it's the call of the Holy?

There are some clues. Is it a call out of your comfort zone and into a new adventure of self-realization? Is it a call to get up and out and do something about the righteous anger you feel? When you cry out, "Somebody should do something!" does the echo come back, "How about you?" Does the voice call you toward your true north, or does it beckon you toward the self-serving, deadly south?

Our cetacean brethren constantly monitor the world around them. They are hyperaware and vigilant. This unwavering and penetrating awareness is essential for their survival and the survival of their world.

How aware are you of the people around you, of the state of the world, of your own unexamined beliefs and assumptions?

When my mind is still and alone
    with the beating of my heart,
I know how much life has given me:
The history of the race, friends, and family,
The opportunity to work,
    the chance to build myself.
Then wells within me the urge
    to live more abundantly,
With greater trust and joy,
With more profound seriousness
    and earnest service,
And yet more calmly at the heart of life.

Paul Beattie, "When My Mind is Still"

We've grown accustomed to closing our eyes to those in need around us. We build highways around certain parts of town. We don't care about what's happening in our neighborhoods, much less in far-off countries that do not affect our national economy or security. In the words of confession from a prayer in a traditional service of Holy Communion:

> Merciful God, we confess that we have rebelled against your love, we have not loved our neighbors, and we have not heard the cry of the needy . . . [10]

Find yourself a good Eli, or a community that can serve collectively as an Eli for you, to help you recognize and respond to the voice of God in your life. Listen up. Listen to see. Listen to get a reliable fix on your present location and chart a course for your desired destination. Listen and enjoy the journey. Listen carefully to discern the voices that beckon you to live and to enliven the lives of others.

The Reflecting Pool

The Reflecting Pool

Quotes to Ponder

If you do not change direction, you may end up where you are heading.

Lao Tzu

Where you are headed is more important than how fast you are going. Rather than always focusing on what is urgent, learn to focus on what is really important.

Stephen R. Covey, First Things First

Questions to Consider

1. Do you have a reliable fix on where you are right now, spiritually, emotionally, relationally, and vocationally?

2. Are you caught in any false feedback loops created by the echoes of your past, the static of your present, or the lack of clarity about your future in which you find yourself to be defeated, depressed, guilty, or immobilized?

Actions to Take

1. Go to a relatively quiet place and attempt to be still. Recognize and name in your mind all of the sounds you hear in one minute. Now, knowing that those sounds are out there, enter into another minute of silence. What do you hear this time, besides those ever-present noises?

2. Draw a picture of your life compass. Identify and name your true north. What is its direct opposite? Write that in at your false south point. Name the relative points between the northeast and northwest, the southeast and the southwest points. What are the due east and due west points in your life, those directions that could easily drift north toward life or south toward death if you aren't careful?

3. Find a reliable sounding board to help distinguish between false feedback and reliable information. Act on your new awareness.

Biblical Wisdom

Be still, and know that I am God. . . .

Psalm 46:10

For we walk by faith, not by sight

2 Corinthians 5:7

A Breath Prayer

Just sit still, breathe, and listen.

Chapter 5

Breathe

Let everything that breathes praise the Lord!
Psalm 150:6

. . . and [God] breathed . . .
Genesis 2:7

Breathing is a conscious activity for cetaceans. It is not an autonomic, automatic, unconscious process for them as it is for us and for most mammals. Therefore, they cannot enter into what humans understand as unconscious sleep. Studies of dolphins have revealed that they shut down half of their brain during sleep. The other half of the brain stays awake to signal when to rise to the surface to breathe and to watch for predators and obstacles. Large whales appear to surface-sleep. Floating horizontally just below the water's surface, they move their flukes periodically to rise above the water for a breath.[1] If cetaceans don't think about it and breathe on purpose, they will drown.

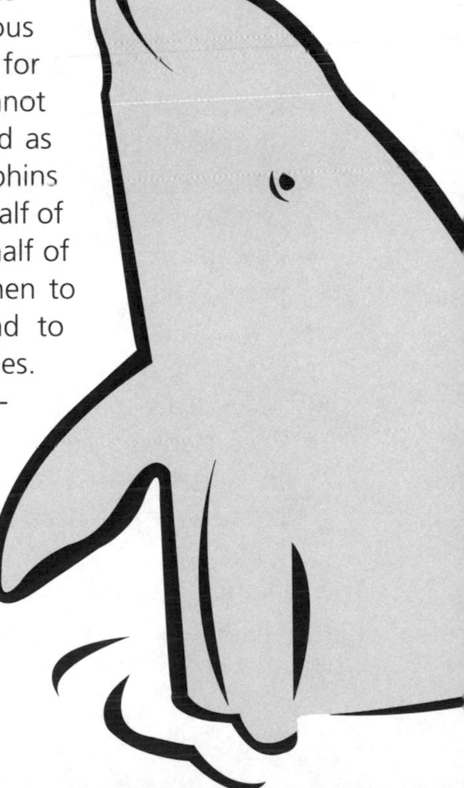

Air Supply

All the cells in your body require oxygen. Without it, they couldn't move, connect, reproduce, and turn food into energy. Without oxygen, all of your cells would die, and you with them! And where does this oxygen come from? From plants, of course. You know that from elementary science lessons. It's a miraculous trade-off. Plants use carbon dioxide, which we can't use, and in fact is harmful to us, to make what we most need—oxygen.

We are surrounded by trees and bushes and vines of every variety, but consider this: the largest biomass on the planet is the global mass of plankton. Plankton produces nearly half of earth's oxygen supply! Phytoplankton produces more oxygen than all other plant life on earth combined and is vital in maintaining the earth's atmosphere.[2] Apart from bacteria, planktonic organisms are the most abundant life form on earth and play a crucial role in the marine food chain:

- Plankton are important in both marine and fresh waters as a vital part of all food webs.
- Phytoplankton are the world's number one source of oxygen.
- Phytoplankton produce about 90 percent of all photosynthetic processes that take place on earth.
- Phytoplankton's key ecological role is the primary production of biomass and oxygen.
- Phytoplankton can absorb marine pollutants (pesticides, mercury) that are carried up the food chain and accumulate to lethal levels in larger animals.
- Phytoplankton are the only plants in the sunlit zone of the open ocean and are thus the major primary producer in the oceanic food web.[3]

The weight of all the plankton in the oceans is greater than that of all the dolphins, fish, and whales put together. Amazing when you think that many plankton are microscopic in size.[4]

Talk about the power of the tiny united in something mighty!

We owe a great debt to our largest cetacean relations, the whales. Heathcote Williams calls them "power-houses in charge of the world's breath."[5] You see, whales are the earth's über-farmers, harvesting gigantic amounts of plankton as they feed, keeping its mass in delicate balance. If there is too much plankton (caused by a decrease in the marine life that feeds on it through our over-harvesting or disease), the ocean's temperature would rise, ever so slightly, yet enough to change world climate, over-heating the water enough to kill off the plankton and with it all living things on earth. The consequences of too little plankton, destroyed by man-made pollutants, are the same.

Breathing 101

Oxygen is only 20 percent of the air we breathe. When you inhale, you inhale carbon dioxide, nitrogen, and oxygen, along with dust, pollen, germs, and other things (like pepper or fumes from chicken farms or paper plants). When you exhale, you exhale some oxygen and all carbon dioxide, nitrogen, and some of these other foreign substances that your body needs to get rid of. You sneeze and cough because of the body's need to expel these unwanted, unneeded, and potentially harmful things from your nose and lungs.

Once inside the lungs, molecules of oxygen and carbon dioxide are passively exchanged by diffusion between the gaseous environment and the blood. Thus, the respiratory system facilitates oxygenation of the blood with the simultaneous removal of carbon dioxide and other wastes. When you breathe you suck in

> No prayer is complete without presence.
>
> *Mevlana Rumi*

more than just oxygen, although that is the only gas you actually need. The body removes 70 percent of its waste products through breathing. Therefore, in the simple act of breathing we receive the most important chemical in our bodies and we exhale what we don't need.

Our noses are pollution controllers. They are designed with a set of filters for clearing the larger particles of dust from the air. Our mouths are not equipped for filtering air. What better reason to keep them closed more often than not!

Sometimes I feel like we've lost our ability to filter out the junk in our culture. We need to develop better personal filters to help us inhale less hatred, fear, and unexamined assumptions.

When we systematically net tons of marine life searching for what we call seafood, we collect tons of species and debris that we don't use or need in the process. This is called "bycatch." For example, feeding dolphins and porpoises are caught in tuna nets only to be maimed or destroyed in the process. We need to develop better filters and systems in our interconnected ecosystem of the earth to help us net what is needed while we release what other species need, as in the process of sharing the atmosphere. We can do a better job of ensuring a "net" gain for all of us.

So you don't think we are interconnected, all sharing the same planetary resources? Go ahead, take a breath. Now, hold on to your part of it. That's it. Don't let it out into the sea of air that circulates around us. Hold on to your air. When you're done, read on . . .

> Fear less, hope more. Eat less, chew more. Whine less, breathe more. Talk less, say more. Love more, and all good things will be yours.
>
> *Swedish proverb*

The average breath contains ten sextillion atoms. Every breath contains approximately one quadrillion atoms breathed by the rest of the world in the past few weeks. Approximately one million of those atoms were breathed personally sometime by every person on earth![6]

Breathe Deep

One of the body's automatic reactions to stress is rapid, shallow breathing, yet proper breathing is the antidote to stress. Breathing slowly and deeply is one of the ways you can turn off your stress reaction and turn on your relaxation response. Most of the time we use only about 20 percent of our lung capacity in shallow breathing. Cetaceans use up to 80 percent. The many benefits of deep breathing include:

- Stress relief;
- Relief from headaches, backaches, and stomachaches;
- Decrease in blood pressure;
- Relief from anxiety and/or panic attacks; and
- The release of natural positive mood enhancers (endorphins) into the bloodstream.[7]

You can use deep breathing anytime, anywhere. It involves no drugs, no doctor visits, and no special equipment. Just do with purpose what you do naturally. You can increase the efficiency of your lungs a thousandfold by increasing your air intake by 5 percent on each breath.[8]

The Breath of God

Rob Bell, in his video meditation called "Breathe" (part of his *Nooma* video series) suggests that the very name of God as written by the Hebrews may simply be the sound of breathing. The word we transliterate as "Yahweh" is really spelled "YHWH" with no vowels (as is the case with all ancient

Hebrew). We'd pronounce the letters: "Yode." "Hay." "Vahv." "Hay." Try it. Pronounce each letter, exhaling with emphasis. To paraphrase Bell, perhaps every time we breathe we are calling the name of God, honoring God, and thanking God for the gift of life's breath.[9]

> Breathe. Let go. And remind yourself that this very moment is the only one you know you have for sure.
>
> *Oprah Winfrey*

Although we don't have to breathe on purpose (except in those rare occasions of awe, surprise, or panic when we forget and someone reminds us to do it), we can choose to breathe *with* purpose. Be aware of your breath. Be aware of the constant life-giving presence of God. Be aware of the holiness around you. Make every breath a prayer, and in so doing, live life as a prayer.

When I ask groups of young confirmation students where God is, most say, "Here, all around us." When I ask them when God is present, they say, "Always." Always and everywhere, huh?

Then I ask, when are you most conscious of God's presence? Many say, "At church or in a worship service." Others say, "In times of extreme pain or ecstasy." Still others say, "When I am out in nature; by a body of water, or on a hike, or at sunrise or sunset." And others say, "At the birth of a baby or the death of a loved one."

Fewer say, "With every breath and every step, every moment and in every circumstance." I want to get there.

It takes practice. Brother Lawrence, a lay member of the Carmelite order in seventeenth-century France, called it "practicing the presence of God."[10]

In 1 Thessalonians 5:17, the Bible encourages us to "pray without ceasing" (KJV). What is the one thing we do without ceasing? Breathe! Unless we do this, we pretty much stop doing everything else. Breathing is the only thing we do without ceasing. So, I guess every breath is a prayer—a prayer of life, a prayer of thanksgiving, a prayer of the eternal now. Of course, this too may be unconscious or conscious. The Bible calls the conscious kind of breathing-as-prayer, "mindfulness." Breathing regularly is not a conscious issue for us as it is with cetaceans. Breathing with consciousness is.

"Pray without ceasing" must mean let every breath be a prayer. First, a prayer of thanks for the breath of life. Where there is life, there is possibility and hope. Second, for the awareness that all of life is a shared, interdependent enterprise. Just try to keep the part of the air that is yours (it's called holding your breath)! Third, for all of the blessings of life, received as gift. And fourth, for the opportunity to be a blessing because you have been blessed.

Pray on purpose—so you can live, and so others may live. *Ohana* is the contemporary Hawaiian word for "family," but its older meaning was "people who breathe together." *Haoles*, which is Hawaiian for "people without breath," was the name given to missionaries when they came to the Hawaiian Islands. Furthermore, "Blessed are the poor in spirit . . . " is the common translation of Matthew 5:3. According to scholar Neal Douglas, in its original Aramaic language, it meant "Happy and aligned with the One are those who find their home in the breathing."[11]

> Live each season as it passes; breathe the air, drink the drink, taste the fruit, and resign yourself to the influences of each.
>
> *Henry David Thoreau*

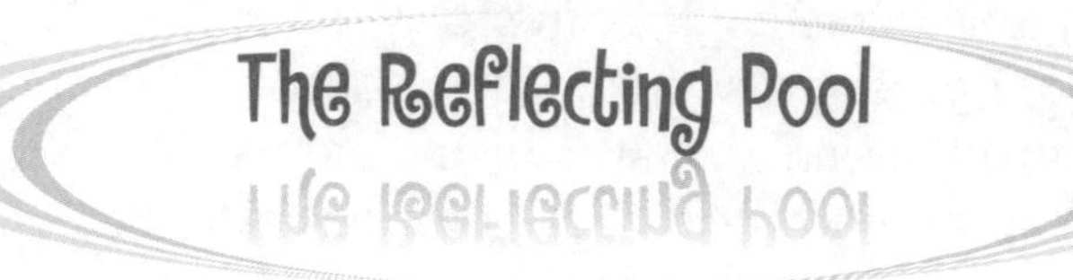

Quotes to Ponder

In every breath I receive everything I need for life in that moment, and I exchange it for everything I don't need.

Howard Hanger, pastor of The Jubilee Community, Asheville, North Carolina

If the only prayer you ever say in your entire life is "Thank you," it will suffice.

Meister Eckart

Questions to Consider

1. What else would you do with more intention or appreciate more if it, like breathing, wasn't automatic?

2. Ever get the "I'll just dies?" You know, "If such-and-such a thing doesn't happen, I'll just die!" Or, "If I can't have such-and-such, I'll just die!" What about the people who really do die because they can't breathe clean air or drink clean water, or buy food or clothing?

3. What if life was worship and breathing was prayer?

Actions to Take

1. Become aware of your normal breathing patterns. Sit quietly and count to yourself slowly as you inhale. Pause for one count when you finish breathing in.

Then exhale slowly, counting to yourself until you have emptied your lungs. Repeat the cycle until you arrive at a regular pattern that feels natural for you. Consciously increase both the length (number of counts) of your inhale and the amount of air you take in a single breath. Breathe more deeply. Then, replace counting with simple, repetitive prayers that take a single breath to pray.

2. Ponder the interconnectedness of all living things and the fact we share all of the earth's available air all the time.

3. Practice the presence of God. Consciously increase your awareness of the Holy in the everyday tasks of life, even each breath.

Biblical Wisdom

For the fate of the sons of men and the fate of beasts is the same. As one dies so dies the other; indeed they all have the same breath and there is no advantage for man over the beasts.

Ecclesiastes 3:19 (NASB)

A Breath Prayer

Thank you.

Chapter 6

Cooperate

Now there are varieties of gifts, but the same Spirit; and there are varieties of service, but the same Lord; and there are varieties of working, but it is the same God who inspires them all in every one . . . he body is one and has many members, and all the members of the body, though many, are one body. . . .
1 Corinthians 12:4–6, 12

There are at least three things we can learn from the metaphor of planet earth as Noah's Ark. One, we are all in the same boat; two, don't destroy the boat; and three, don't miss the boat and do everything in your power to make sure others don't miss, or mess with, the boat.

You can't inhale and then hold onto "your" air. The young boy, Eli, reminded us (in chapter 3) that we all share the "same" water. And there is only one inhabitable planet in our neck of the universe with a finite amount of natural resources. Like I said, we are all in the same boat.

It's wise to consider and value what we have in common. It amounts to more than the ways we seek to make ourselves, our cultures, and our nations

"different." Don't get me wrong. I marvel at the uniqueness in the fingerprints and retinas of each individual human being ever born. I appreciate the foods, music, art, architecture, and sensibilities of the wondrous mosaic of cultures on this Planet Ark. And yet, the need for air, food, water, companionship, safety, shelter, dignity, and aspiration are our common foundation that runs beneath and through the varieties of culture. The mystics of all faiths recognize that what is true about any person at the deepest level is true of all persons. There already exists a unity in diversity. Our choice is to embrace it and make it a global community value.

There's Only "Us"

We are all in this together. We can no longer afford the luxury of national civil wars and international pissing contests. "Think globally and act locally" is a prophetic slogan that has come into its own. The fact is, when we act (or fail to act) locally, we have a global effect. The effect of the local to affect the global is everywhere apparent. Local economies affect the global economy; local ecologies affect global ecology; and local political upheavals and natural disasters affect global consciousness and responsiveness.

Perhaps the biggest barrier to realizing a comprehensive inclusivity and a global ethic of interdependence are the ingrained dichotomies embedded in the rock logic of our collective memories:

- Sacred vs. Secular
- Learner vs. Teacher
- In vs. Out
- Body vs. Spirit
- Life vs. Work
- Male vs. Female
- Allies vs. Enemies
- Conservative vs. Liberal
- Leader vs. Follower
- Church vs. Churches
- Pastor vs. People
- Boss vs. Employee
- Gay vs. Straight
- Lay vs. Clergy
- Religion vs. Spirituality
- Top vs. Bottom
- Sinner vs. Saint
- Us vs. Them
- Master vs. Servant
- Nation vs. Planet
- *Kairos* (Eternal Time) vs. *Chronos* (Clock Time)

Historic hatreds and viral predispositions toward mistrust and fear of others are perpetuated by divisive dualisms of race, gender, sexual orientation, age, region, culture, religious tradition, perspectives, philosophies, and worldviews. The old rock logic of either/or thinking is hard to break. Remember, water is receptive and seeks to fit and flow.

Remember what Jenny Cannon wrote (in chapter 3), "Water forms around us, shapes around our bodies, and when we share water space with someone it makes room for us all." Water will adapt to the contours of its environment. Rocks cannot and will not adapt to or make any attempt to fit the container they are put in. You make concessions to them on their terms or they will bruise you, cut you, and punish you.

Rock logic is at the heart of what we have come to call the Golden Rule: "Do unto others as you would have them do unto you." In this dictum, your preferences are the standard for your treatment of others. Unlike at Burger King, others must have it your way, not theirs.

Water logic supports the practice of what futurist/historian Leonard Sweet calls the "Platinum Rule" of Christianity, "Do unto others as *they* would have you do unto them."[1]

Jenny Cannon's epistle speaks of a generous hospitality in the metaphor of water that "forms around us" and calls us to "form around others" as a water-community of witness and care (Heb. 12:1), practicing our faith by Sweet's "Titanium Rule": "Do unto others as Christ has done unto you."[2]

Another barrier to a global culture of interdependence is aggressive nationalism. It's easy to understand the nationalistic spirit, historically hard-fought and won with terrible human sacrifice. But this is a time that calls for a humble patriotism that comes with the awareness that many countries were established by conquerors who took the land from its original inhabitants. Many a wrong has been done in the name of the victor's manifest destiny. From space, the big, blue marble of Earth looks like one big, living organism. There are no pastel-colored countries with dotted borders like the globes in libraries and classrooms.

The song "From a Distance," written by Julie Gold and recorded by Nanci Griffith and Bette Midler, affirms a global perspective:

From a distance the world looks blue and green,
and the snow-capped mountains white.
From a distance the ocean meets the stream,
and the eagle takes to flight.

From a distance, there is harmony,
and it echoes through the land.
It's the voice of hope, it's the voice of peace,
it's the voice of every man.[3]

The chorus isn't so helpful: "God is watching us *from a distance*." This is poor theology if one experiences the Holy as One who is with us. God's most repeated promise is "I will be with you." Remember the name for Jesus that affirms an intimate God, and immanent God—Immanuel—means "God with us!" It's this kind of theology that nudges us toward intimacy with our neighbors in widening circles across boundaries of country, creed, and color. We can see the reality of our interdependence from a distance. It's way past time to see it up close and personal.

It's a Both/And World for Us

Remember the world-changing question and answer of quantum physics, "Does light behave like a wave or a solid particle?" Answer: "Yes." Is Jesus fully human or fully divine? Yes. Are we living or are we dying? Yes. Are we citizens of a country or citizens of the planet? Yes. Are we citizens of the planet or citizens of the Kingdom of God? Yes. It's time to embrace the "both/and" and examine more closely our attachment to "either/or" thinking or its close relative, the "bottom line." Maybe that will free us to ask, "What is the best for the most?" or "Might there be a third way?"

Cetaceans have developed a dual system of blood flow that enables them to regulate their body temperature by either conserving or dissipating heat, depending on the circumstances. If a whale's body temperature is too high, this countercurrent blood exchange allows for blood circulating in the outer skin to release its heat into the environment. If the whale is too cold, blood is returned to the heart through a different set of vessels that are wrapped around warmer interior arteries, heating the blood to raise the creature's temperature. The evolutionary process did not choose one blood flow system over another. Both are necessary for the sea mammals to function to the peak of their potential.[4]

> Interdependence is a fact, it's not an opinion.
>
> *Peter Coyote*

Sometimes the questions we ask imply that there is only one answer out of limited options. A good example is, "Is your glass half empty or half full?" The question appears to present only two possible answers with an added bit of pressure to be optimistic. But there is a third answer that is more gracious, more real, more authentic. The glass is always *full*! The glass is full of some percentage of water and some percentage of air—maybe half, maybe more or less, both necessary for our survival. The glass is always full of some joy and some sorrow, some clarity and some ambiguity, some sense of something and some sense of nothing, but always full. The beauty of cetaceans is that their very existence insists that we abandon our simplistic either/or thinking, at least in some cases. To ask a cetacean if it is a mammal or a fish is the wrong question, unless you can live with the answer "yes."

In-body-ment and In-corps-oration

"Now you are the body of Christ and individually members of it" (1 Cor. 12:27). This is the way Paul describes the ideal embodiment of true community. The church envisioned as the body of Christ incorporates many of the values and perspectives of the "futurepresent" world: connectivity, intimate interdependence, a systemic sensibility, and incarnational theology and missiology. The "body" passages, particularly in the New Testament, describe a profound interconnection and interdependence.

We dare not approach the body from a mechanistic "parts" perspective lest we lose the systemic wholeness of it. We cannot disregard or marginalize one part in favor of another; in fact, we invest more in the less-presentable parts. Interdependence is not based on everybody in a system doing the same thing the same way. It is everybody finding their way to support the same goal or mission within a living system; like your discrete bodily tissue groups and systems, in concert contributing to the same mission—namely, your life. "For as in one body we have many members, and all the members do not have the same function, so we, though many, are one body in Christ, and individually members one of another" (Rom. 12:4–5).

In this liberated gospel, we are free to honor God in the temple of our bodies and not despise them. In the post-postmodern church:

> We will see ourselves [and others] as far more than minds (or spirits) that temporarily need bodies (mere throw away containers, an embarrassing concession to the material world) to get around. Instead we will grapple again with the ancient Jewish insight that when God made us, God made us as bodies, and pronounced those bodies "very good."[5]

In the progress of personality, first comes a declaration of independence, then a recognition of interdependence.

Henry van Dyke

"And the Word became flesh and dwelt among us . . . " (John 1:14). In the Incarnation, God is in-*corps*-orated and in-*bodied*. Jesus " . . . shared the human frame, and for all human beings, his body was the focal point of his life."[6]

In the continuing incarnation, we are invited, even called, to in-body this new inclusive, interdependent gospel. The new authenticity comes from embodied faith, that is, doing what you profess, not from titles, office, or credentials. "Now you are the body of Christ" (1 Cor. 12:27).

When Christians take communion, we affirm that we become what we eat, namely the Body of Christ in the world. The communities of Christ understand themselves to be part of the continuing Incarnation of Christ as part of the "holy catholic church," the universal body of Christ, the continuing incarnation of Jesus in the flesh and blood of his disciples.

Mike Holly put it this way:

> . . . if we are the second incarnation; if we are the living body of Christ, then the way we live, the way we treat each other, the way that we form our churches is a witness to Christianity and to Christ himself. So, I mean, if we're stuck in modes that aren't doing justice and loving kindness and loving people, then we aren't being faithful witnesses.[7]

We are also part of the body of all creation, interconnected and interdependent. There is one race and it is human; one family of *adam/adama*, and in fact, one family of all creatures. "The earth is the LORD's and everything in it" (Ps. 24:1 NIV). We affirm with the prophets of this Ecozoic Era that "the universe is a communion of subjects rather than a collection of objects. Existence itself is derived from and sustained by this intimacy with each being with every other being of the universe."[8] Remember "God so loved the world . . . " (John 3:16).

Unlike normal photographs, every small fragment of a piece of holographic film contains the information recorded in the whole.[9] In our bodies, each tissue group may have different functions and vastly different physical qualities, yet the whole of our DNA identity is included in every cell. Likewise, the body of creation, in all its individual members, all of its species, and in its wholeness, is a hologram of terrestrial life. Corporately and individually, we share the same building blocks of life. The Cherokee use the phrase "Ho! Mitakuye Oyasin," meaning, "We are all related," as a greeting and a blessing at the end of prayers—a reminder that we are kin to the four-leggeds, and those who fly, swim, and crawl upon the earth.

Heads Up

A metaphor that needs some reframing for our age is that of "the head." The premodern understanding of the head was that it is the crowning glory, the literal top of the heap in the hierarchy of the body. In Colossians 1:18, Jesus is described as "the head of the body, the church" (KJV). Jesus is the head, "from whom the whole body, nourished and knit together through its joints and ligaments, grows with a growth that is from God" (Col. 2:19 RSV).

Following that logic, the lower extremities would be the literal lowest of the low. Fortunately, Jesus' upside-down wisdom puts the last first and the prophet blesses "the feet of the one who brings good tidings, who publishes peace . . . who publishes salvation . . . " (Isa. 52:7 RSV). And that blessing extends to our hands, those extremities with the opposable thumbs that create, heal, express, love, touch, feel, offer, and support.

The modern perspective of the head intensifies the hierarchical model of premoderns and adds the bureaucratic dimension of the CEO—the boss or executive director that ultimately calls the shots on behalf of the body. This view of the head and outdated approaches to brain science and organizational effectiveness perpetuate a paternalistic hierarchy.

A perspective of the head as an interface or nexus of information and a part of the whole is more useful and applicable in our time. This perspective supports an interactive, inclusive, flat organization in which more voices can be heard and participation can expand.

Community as Wayang Kulit

Television gives the appearance of being a part of a crowd without the benefit of actually interacting with fellow audience members or experiencing the dynamics of being in the midst of a live audience. Traditional Western theater is predicated on a kind of unspoken agreement between the audience and the creators of theatrical illusion that keeps the audience in the dark—as passive observers of the presentation.

In his book *Interactive Excellence*, Edwin Schlossberg observes, "Except in the case of highly experimental theater, any audience awareness of the process that creates the illusion has usually been considered a failure of that process."[10]

Both of these forms of communication are at best one way, a telling. In television, interaction and conversation with the presenters is near impossible. There is no dialogue, only sales pitches. In theater, interaction with the players or conversation with fellow audience members is, at best, considered rude.

By comparison, Indonesian shadow-puppet theater, *wayang kulit*, is truly a shared community experience. Since the lengthy plays are always based on familiar stories, the audience can concentrate on comparing it to previous performances and exploring all facets of its production and execution, literally from all angles and aspects, during its presentation. Schlossberg describes his experience:

> One part of the audience chose to step into the illusion, listening and watching from the front, while the other part chose to be with the performer, to explore the art of his presentation. This practice, so alien to Western tradition, allows the audience to learn how something is made and how well it can be done. Their understanding, discussion, and appreciation of all facets of the play are part of the presentation. The culture values the audience's active role in the process as equal in importance to that of the puppeteer or musician.[11]

Schlossberg, a developer of interactive museums and exhibits, asserts that the way to develop better art is to develop better audiences. "Better" here means knowledgeable, experienced, well versed, and fluent so that real dialogue, participation, and interaction with the creation, the creator(s), and fellow audience members or colearners can take place. A truly interactive experience is one in which a participant's involvement affects the outcome.[12]

Talk about a stake in what happens in the world around you! The expanding access to shared information and the emerging culture of interdependence will enable people to move from merely receiving controlled and managed presentations to engagement in open and informed participations. In a philosophy reverse of *The Wizard of Oz,* we will encourage each other to pay attention to the one behind the curtain. In institutional church culture, lay people will touch holy things!

Sweet points out that Jesus' restriction on touching him after his resurrection no longer applies; the condition that he return to his Father has been met. Touch away![13] "The cup of blessing which we bless, is it not a participation in the blood of Christ? The bread which we break, is it not a participation in the body of Christ?" (1 Cor. 10:16). A culture of interdependence demands this same kind of access to hands-on participation in all kinds of organizations and political processes that can only be effective with access to shared information.

The dynamics of our interrelationship, interdependence, and connectivity are readily apparent in nature as we examine ecosystems and the impact of any action in one area on the others. We are connected not only organizationally or because of our global economy, but also at much deeper levels, one of which is ecologically, as part of that body of the Creation.

The Internet works because of interconnectivity, fueled by access, described by David Weinberger as "small pieces loosely joined."[14] It is a huge unregulated network that continues to grow exponentially, not because it always runs smoothly or because everybody in the expanding Web has a great deal of expertise in digital communications technologies, but because it is millions, soon billions, of people doing the same simple process in search of something they find valuable—information they can use and connection to others (passion and people).

> Interdependence is and ought to be as much the ideal of man as self-sufficiency. Man is a social being.
>
> *Mohandas Gandhi*

Small pieces pointing to other small pieces. This sounds a lot like the classic definition of evangelism, "one beggar showing another beggar where the bread is." Author David Weinberger puts it this way: "We're getting to know people in many more associations than the physics of the real world permits."[15]

Non-linear interconnectedness is the key to rapidly expanding networks across all barriers of time, place, economies, and cultures. It's about connecting people in emergent communities of passionate purpose—connecting in relationships, connecting in mission, connecting in global consciousness and partnership, connecting in local embodiment and action. It's about connecting beyond *our* business, *our* country, or *our* people. Weinberger says, "Sad and dying is the insular [community or organization.] Limited gene pools produce woeful results over time . . . "[16]

Emergent Communities of Passionate Purpose

Emergent behavior can be described as leaderless, self-organizing, coordinated group behavior in which the whole is greater than the sum of its parts. Rather than starting with a plan, we start with a reason and develop and adapt the plan as we go along (like building a plane as we fly it). This can only happen when the sharing of information within the system is maximized. Open information generates positive [co-creative] energy. This chaordic dynamic births temporary communities of passionate purpose, spontaneous aggregations or "bodies" of people around a shared motivating mission.

It's like slime mold. Yes, slime mold.

> The slime mold spends much of its life as thousands of distinct single-celled units, each moving separately from its other comrades. Under the right conditions, those myriad cells will coalesce again into a single organism, which then begins its leisurely crawl across the garden floor, consuming rotting leaves and wood as it moves about . . . [all the "its"] become a "they." The slime mold oscillates between being a single creature and a swarm.[17]

How do the simple, cellular units know when it's time to group up? Who organizes them and charts their course? What motivates their aggregation? What is the key to their morphogenesis?

Conventional wisdom holds that a call or signal of some sort is put out by some superior, designated leader; perhaps a sort of queen bee, or the group's pacemaker cell. In the case of slime mold, they follow a trail of a signal substance, called acrasin or cyclic AMP, thought to be emitted by a few elite leader cells. Stephen Johnson observes, "And for millennia we've built elaborate pacemaker cells into our social organizations, whether they come in the form of [monarchs], dictators, or council [members]."[18]

Johnson says we've built our modern command and control systems and hierarchies on the model of queen bees and pacemaker cells. But, surprise! There is no cellular hierarchy among slime mold. All "slime mold cells are created equal. . . . The slime mold cells were organizing themselves."[19] Yes, they were following the scent of cyclic AMP, but individual slime mold "trigger aggregation . . . simply by altering the amount of cyclic AMP they released individually . . . based on its own local assessment of the general conditions." In this way, "the larger slime mold community might be able to aggregate based on global changes in the environment—all without a pacemaker cell [or elite category of cells] calling the shots."[20] Action begets action.

Follow the gospel according to Nike, "Just do it!" and others will join you, and still others will join them. For the Christian community, our AMP is "Actualized Missional Passion." Its fuel is active compassion. The more active compassion we embody, the more AMP we secrete, and the more people will join the movement.

These temporary communities of passionate purpose are classic bottom-up systems; simple agents doing simple things in aggregation that leads to more complex behavior that is greater than the sum of their parts. "Temporary" may mean a few minutes, or it could mean decades. It could result in an institutionalizing cycle and the creation of bureaucracies and structures (ostensibly to support the mission), but most will not. In fact, part of their DNA is a kind of "rock blocker"—a trigger to motivate the group to morph or dissolve as conditions change in a fluid monitor-and-adjust mode rather than a rigid control-and-perpetuate posture.

Leaderless Initiatives

These leaderless initiatives are led or driven by the mission, not by a designated leader or a boss. All partners in the network are reading the signs and are in constant communication with each other by creating creative feedback loops. Anyone can initiate movement, action, the sequence of events, push the start button, or put out the call. No permission is

> The life I touch for good or ill will touch another life, and that in turn another, until who knows where the trembling stops or in what far place my touch will be felt.
>
> *Frederick Buechner*

needed or sought. No vote is taken. If there is an aggregation, a gathering, a critical mass formed around the task at hand, so be it. If not, so be it.

That's the way of self-organizing groups. When the aggregation develops, there is a pooling of resources, each one offering and giving what they have in personal gifts and skill sets, as well as physical and fiscal resources. Pooled resources flow to the mission. Again, constant communication helps leadership emerge based on giftedness and experience among co-missioners before, during, and after the mission at hand, which is part of a larger matrix of mission (life, survival, community, creation, and recreation). When the mission at hand is complete, all share success and/or failure, joy and pain, gains and losses.

> The new electronic interdependence recreates the world in the image of a global village.
>
> *Marshall McLuhan*

Examples of emergent, decentralized, loosely joined, temporary, mostly small communities of passionate purpose include the following:

Food, Not Bombs

This is a grassroots feeding program for urban street communities. There is no central office or governing board. But there is information and story-sharing on the World Wide Web and a growing network of participant/initiators who share their process as they travel and converse with others across the globe. Local operations consist of collecting surplus produce, day-old bread, and other foodstuffs that may be past the official sell-by date; preparation of simple recipes and spreading out raw foods and produce in a community area and offering it to all who are hungry and thirsty. Community building among community members and transients, caregiving, and celebration are main courses at the "table."

Appalachian Music Community Gatherings

Without a central committee or governing board, lovers of Appalachian music gather local and regional music festivals. Anyone can put out the call for a jam session, and if conditions are right, community support, food, music, and lodging all materialize, stone-soup style.

Fish Hauling in the Abacos, Bahamas

Gathering bait fish is a regular feature of the cyclical life of the island community. Everyone knows it's going to happen. They just don't know exactly when. When an uninitiated visitor asks, "Will we go fish hauling tomorrow?" the response comes back, "Won't know 'til morning." All stakeholders in the process are reading the signs, and anyone can put out the call. Some have boats; one has the best net; a few have history and knowledge of the best places to find the cigar-shaped ballyhoo; some know the technique for getting the highest yield with the least loss in hauling in the net; some have muscle power; and some have only delight, curiosity, and rookie enthusiasm to share.

All give what they can, and all get what they came for: a half-season of bait fish for the commercial fishermen, inexpensive retail fish for the owner of the local store, fish for the grill for the impromptu team, and the experience of a lifetime for eager tourists. And everyone receives gifts of community bonding, cooperative hard work, and lots of fun. No committee, no government regulations, no designated leader—just a shared mission and the offering of each person's unique gifts to the process.

There are limits to purely emergent communities of passionate purpose. Some processes and tasks take a greater amount of intentional planning and organization to be effective. My friend Ed Kilbourne points out that a group of lay people can't just decide to perform brain surgery or a heart transplant. Pastor and teacher James Howell observes that organizations often develop from simple beginnings because of the need to expand services, deploy resources, and accomplish more than could be possible without some intentional structure.[21]

You Can't Force Emergence

> You will never get anywhere with a dolphin by force. If you try it, he'll break off contact, retire to the furthest corner of the pool and ignore you. If you persist, he will go on a hunger strike. He'll let himself die rather than submit to doing something against his will.[22]

In what George Hunter, author of *The Celtic Way of Evangelism*, calls the "Roman Way," the way to effect change is to disregard and overpower the old and enforce the new.[23] We see this model played out in military coups and imposed democracies. We have seen this model played out in historic forms of evangelism, as Western European conquerors imposed their culture and religion on indigenous peoples.[24]

In what might be called the "Bureaucratic Way," change comes only after years of planning, passage through approved channels, and implementation on a strict schedule by designated authorities.

Jesus embodies another, more invitational and incarnational, way to effect change. His approach begins with a commitment to live among and learn about a people's worldview, its deepest desires, and its greatest fears. He casts visions of community and interdependence, valuing each person's unique contribution to something bigger than themselves. His is a way of cooperation in community.

He begins by proclaiming the good news of God's preferred vision for creation and all of its creatures. He uses images that everyone can understand and bring a bit of their own perspective to. His gospel banishes false and oppressive dichotomies, and he cultivates a bottom-up inclusive community. His incarnational way gets people in-*corps*-orated—included into the body (*corpus*).

The earth is a corporate body, a living organism with many interdependent parts. This is our reality. It does not need to be forced. It needs only to be recognized, and we need to cooperate.

Don't miss the boat!

"No Man Is an Island"

by John Donne

No man is an island entire of itself; every man
is a piece of the continent, a part of the main;
if a clod be washed away by the sea, Europe
is the less, as well as if a promontory were, as
well as any manner of thy friends or of thine
own were; any man's death diminishes me,
because I am involved in mankind.
And therefore never send to know for whom
the bell tolls; it tolls for thee.

The Reflecting Pool

Quotes to Ponder

The world is my country, all mankind are my brethren, and to do good is my religion.

Thomas Paine

Humankind did not weave the web of life. We are but strands within it. Whatever we do to the web, we do to ourselves.

Chief Seattle

Questions to Consider

1. What is your response to the phrase, "If you want something done right, do it yourself"?

2. Do you believe that there are some things that you should not know?

3. Can you live with the answer "yes" in response to an either/or question?

Actions to Take

1. Go rafting. Pay attention to when it's best to go with the flow, cooperating with the stream, and when it's best to paddle against it . . . still cooperating, but using its power as leverage to move you where you want to go.

2. Gather some family and friends together and play some cooperative games—games that by definition are about group goals, high energy, and fun instead of defeating someone or the other team. Check out http://www.mrgym.com/CooperativeGames.htm and http://www.inewgames.com/.

3. Trace the food chain back from what you ate for dinner.

4. Play a game of Jenga after labeling all of the pieces: a few as water, a few as clean air, a few as healthy soil, and the rest as different species of animals.

Biblical Wisdom

If the foot would say, "Because I am not a hand, I do not belong to the body," that would not make it any less a part of the body. . . . The eye cannot say to the hand, "I have no need of you," nor again the head to the feet, "I have no need of you." . . . If one member suffers, all suffer together; if one member is honored, all rejoice together.

1 Corinthians 12:15, 21, 26

A Breath Prayer

Place me in the flow of your Spirit, O God.
Help me know when to paddle and when to lift my oars.

Chapter 7

Live

I came that they may have life, and have it more abundantly.
John 10:10

All animals, except man, know that
the principal business of life is to enjoy it.
Samuel Butler

The primary purpose of life is life. Life is its own purpose. Not *has* its purpose, *is* its purpose. We go to great lengths to save it and protect it. In an emergency room, no one is asking, "What purpose does this person serve on earth?" No, everyone is urgently trying all of their practiced techniques and employing all available technology to save a life.

Why do some people do all in their power to save a life? Some people even risk drowning or death by fire to save the life of an animal. Why? Because life itself is precious. It is not primarily its purpose but its fact that makes it so precious. Before we can get to some sense of purpose for our lives that may drive (or preferably draw us) it's essential to grasp this.

Life is persistent. Life begets life. According to Christian mystic Howard Thurman, "At the core of life is a hard purposefulness, a determination to live."[1]

God's purpose may be no more, and certainly no less, than life itself. Jesus says, "I came that they may have life, and have it more abundantly" (John 10:10). More life, not more stuff. The writer of the Gospel of Luke reminds us, "a man's life does not consist in the abundance of his possessions" (Luke 12:15). The gift of God in Jesus Christ is not more time, more achievement, or more false security. Just more life.

> Every blade of grass has its angel that bends over it and whispers, "Grow, grow."
>
> *The Talmud*

One of the golden threads that connect all of Arthur C. Clarke's works of science fact/fiction is the persistence of life, the determination of life itself to continue, develop, and propagate. We see grass pushing its way through asphalt. We wonder at sea creatures living under extreme pressure in the vortexes of boiling geysers of poisonous subterranean water. Flowers bloom and reptiles thrive in the deserts while other species make their homes in the frozen tundra near the earth's poles. And fresh, healthy vegetation grows from the ashes of forests cleansed by fire.

The Creator is generous, patient, and wise in the continuing procreation of life on this planet. The ratio of seeds on the head, or "clock," of a dandelion that actually survives to germinate and become the next generation of seed-producing flowers is less than two in one hundred![2] It's a good thing there are so many seeds in most fruits and vegetables. Some seeds are used for food by other animals, worms, and insects in the food chain. Some never germinate due to adverse weather conditions that may bring too much water or not enough, or winds that are too strong or too weak to effectively spread airborne seeds around. Some seeds never mature due to chromosomal breakup. And some never penetrate the ground.[3]

Only between 5 and 10 percent of the annual frog population reaches adulthood, with 25 percent dying before hatching from their eggs, 25 percent perishing as food for predators, and one-third killed by the elements.[4] There are between 200 and 500 million sperm in the average human male ejaculation, with few surviving the short swim to their intended destination, where only one is needed to fertilize the female egg![5] Apparently, it's not that easy being conceived, let alone making it to adulthood. God is extravagant with the raw materials of life on this planet. Against all odds, with the possible exception of our free will, life persists.

Live. On purpose, as purpose. Celebrate the fact of being. Life is for life. In the words of Paul McCartney, "Let it be" with the accent on *be*. God's name is "I AM WHO I AM," (Exod. 3:14), not "I am what I make," or even "I am what I do." This is the supreme declaration of being.

Grammatically speaking, we don't say, "I be." Yet, "to be" is the root verb of the definitive declarations, "I am," "you are," "they are," "we are." When we realize the profound nature of these foundational truths, we begin to realize that people are not things, and groups of people cannot be dismissed by dehumanizing or disregarding them.

"And are we yet alive?"[6] The hymn writer Charles Wesley more declares than asks. And are not others alive? And is not the planet a living, breathing organism? What shall we do because we are? How shall we live because others are?

Lao Tzu, the old sage and father of Taoism, teaches, "The way to do is to be."[7] Let's begin and proceed from there.

Live, Don't Merely Exist

Live while you are alive. Thurman suggests, "Don't ask what the world needs. Ask what makes you come alive, and go do it. Because what the world needs is people who have come alive."[8]

Live, don't merely exist. To the porpoise, life is lively. Life is the purpose of the life of the porpoise. They seem to get the inside joke of Ecclesiastes: live, love, laugh. Eat, drink, and be merry, enjoying life as gift, not life as test or as punishment. This is the true prosperity gospel—that life itself is the richest gift of all (as compared with the false prosperity gospel proffered and profited from by televangelists who sell the idea that God wants true believers to be financially rich).[9]

> True religion is real living; living with all one's soul, with all one's goodness and righteousness.
>
> *Albert Einstein*

The writer of Ecclesiastes, who goes by the name "The Preacher, Koheleth," may be seen on the surface to be a real fatalistic pessimist! He goes on a quest for meaning in life and comes to the conclusion that it's all futile, empty, and pointless. It's all done in vain. It's all vanity. Not the spending-too-much-time-in-front-of-the-mirror, it's-all-about-me kind of vanity, but vanity in the sense of being meaningless, wasted, fruitless.

Koheleth finds the pursuit of riches and power is vanity because it will all come to nothing when you die. He finds that while you live, the more you have, the more you have to worry about, protect, and maintain. The pursuit of wisdom and knowledge is vanity because it does not help you penetrate the mysteries of God or protect you from any of the calamities that befall all people. Remember, Jesus says, "The rain falls on the just and the unjust" (Matt. 5:45). The pursuit of mere sensual pleasure (such as gluttony, drunkenness, and sexual promiscuity) proved to be meaningless and alienating. None of these pursuits ensure a lasting legacy, the enduring of one's name and work after death. Even service can become self-serving and, therefore, meaningless and vain.

In his more pessimistic moments, Koheleth could be seen as the originator of the infamous bumper sticker: "Life's a b—— (um, a big challenge), and then you die." Koheleth's philosophy has been has been twisted through the years to become a slogan for hedonism: Eat, drink, for tomorrow we die (1 Cor. 15:32 NIV). The Preacher's advice is not an invitation to debauchery. It is not to equate sensual gratification with joy. It is a wise insight to find joy in the moments of life. Don't postpone joy. Celebrate life and all that it brings your way. It is an invitation to declare with the hymn writer, "Whatever my lot, thou has taught me to say, 'It is well, it is well with my soul.'"[10]

According to the Preacher, the only reason to be alive is to enjoy it, and this in itself is pleasing to God. This Preacher is such a realist that his writings had a hard time making it into the canon of the Bible. His perspectives belong with the likes of Job. He doesn't believe in an afterlife, so he's looking for meaning in the here and now. But the Preacher is on to something: The very ability to enjoy the gift of life itself, is itself a gift!

The Preacher would agree with Emily Dickinson, who once remarked, "The mere sense of living is joy enough."[11] According to one commentator on Ecclesiastes, "All that he can say with any degree of assurance is that it is God's will that we live happy, successful, fulfilled lives while we are young."[12]

> Ethics, too, are nothing but reverence for life. This is what gives me the fundamental principle of morality, namely, that good consists in maintaining, promoting, and enhancing life, and that destroying, injuring, and limiting life are evil.
>
> *Albert Schweitzer*
>
> Civilization and Ethics, *1949*

Geoff Klock, philosopher, author, and blogger from Astoria, New York, offers this interpretation of Koheleth's gospel from Neutral Milk Hotel's album, *In the Aeroplane Over the Sea*:

> And one day we will die
> And our ashes will fly from the aeroplane over the sea
> But for now we are young
> Let us play in the sun
> And count every beautiful thing we can see. . . .[13]

The Preacher of Ecclesiastes would have us to be so aware of the joy in each moment that we would not be preoccupied with the struggle and pain, even the seeming futility, of life. That we would find the joy and share the joy of whatever we have and do.

Leonard Sweet asks the question, "Do you think Jesus ever got bored?"[14] It isn't mentioned in the Bible. Sweet suggests that maybe he was so engaged in life that he never had a chance to be bored.

My Hebrew teacher at Duke Divinity School closed each class with the same admonishment that he received from his Hebrew teacher, "Be happy in your work!" It has a certain sadistic quality to it, at least in the ears of struggling linguists. Yet, Koheleth basically says, "Eat, drink, and be happy in your work."

Memes

Much of our value system and unexamined attitudes are perpetuated by strong social viruses called "memes" (rhymes with "dreams"). A meme is to a society, culture, or body politic what a virus is to a physical body and a population. It's an infectious idea that gets passed rapidly from person to person until it becomes an accepted norm, or at least a prevailing convention. Memes run deeper than mere fads. Examples include "the American Dream" and global warming. Over time, these ideas take on a mythic quality as conventional wisdom. They persist until another more persistent or compelling meme supercedes them. Like a virus, it's hard to kill once it takes hold in a body. It has to run its course or be suppressed and superceded by a newer, stronger virus. One meme that has dominated USAmerican culture is the Protestant work ethic, or PWE.

At its root is a theology of "work as a punishment" for rebellion against God, the consequence of the Fall as told in Genesis 3. Work is a duty, an end in itself, whose chief goal is to generate personal wealth. Work is meant to humble the worker by virtue of "doing what you are told." Work is elevated to the most important thing in a compartmentalized hierarchy of life's tasks and roles. Our world is work-centered. We live to work and we work for "retirement"—when real life begins and we are free to do what we want.

> To live is so startling it leaves little time for anything else.
>
> *Emily Dickinson*

In the logic of the PWE meme, life is compartmentalized into dualisms of work and play, work and vacation, work and recreation, time "on" and time "off." The prevailing practice is to lump all of your fun and joy and what you want to do into the weekend (the week being something you want to end, to be over with); Friday night through Sunday night are the days you work for, what you postpone life for.

The PWE deems idleness as the devil's workshop, while the workplace is part of someone else's business. Bosses and managers direct the workers until work becomes so "other directed" that people don't know what to do with unstructured time. Free time is filled with more "busy-ness" or it becomes mere idleness that is as excruciating as forced labor. Free time is even guilt-producing for many. Sunday, a day of Sabbath rest, is only rest from work. Personal time, the remainder, belongs to someone else (the boss or the company).

The Digital Work Ethic Meme

The digital work ethic (DWE) meme is suppressing and superceding the PWE as the new common sense of life and work. It originally grew out of the community of computer hackers to spread into other communities and subcultures. In the same way that the old PWE wasn't about being Protestant or particularly religious, yet became conventional wisdom about human nature and our deepest values, the new DWE transcends any particular role in society to transform values and practices across the social and occupational spectrum. Although it began as a "hacker ethic," everyone who lives the digital lifestyle is now called a "hacker."

If the PWE meme is about living to work and make money, the DWE meme is about living free to create, make a difference, and connect with others in the process. This freedom is a "complete freedom of expression in action, privacy to protect the creation of an individual lifestyle, and a rejection of passive receptiveness in favor of active pursuit of one's passion."[15]

This freedom includes more self-determination in use of time. It fosters a non-compartmentalized life and lifestyle and banishes the old dichotomies of work and play. In new combinations of life's roles and tasks, play, work, family time, and recreation are all part of an animated tapestry rather than separate blocks in a set of Legos. The weekend is now seven days a week. Life is not so much leisure-centered as life-centered.

> Hackers do not organize their lives in terms of a routinized and continuously optimized workday, but in terms of a dynamic flow between creative work and life's other passions, within which there is room for play. The hacker work ethic consists of melding passion with freedom.[16]

For the hacker, and the dolphin, work is play.

Hackers are not motivated by money or institutional validation. They are motivated by freedom, passion, and connection to others who share their passion. In fact:

> . . . passion describes the general tenor of their activity, though the fulfillment of it may not be sheer unmitigated bliss. . . . The hard work and dedication will become a kind of intense play rather than drudgery. Meaning cannot be found in work or leisure but has to arise out of the nature of the activity.[17]

Their three top motivators are passion for what they are creating, the social value of it, and the recognition of their peers and partners in the enterprise. They have no problem with making money if these passions and freedoms are part of the equation.

For hackers:

> . . . concern for others [is] an end in itself and a desire to rid society of the survival mentality [and exclusiveness] created by its [rock] logic. This includes the goal of getting everybody to participate in the network and to benefit from it [accessibility and participation], to feel responsible [rather than expecting someone else to be responsible] for the long-term consequences of a network society, and to directly help those who are left on the margins of survival.[18]

This could be a quote from a Christian mission manual!

Related Expectations

The inhabitants of the digital world have different working definitions and experiences of time and space, freed from conventional assumptions about either.

> Ultimately, matter doesn't matter. If we can be together so successfully in a world that has no atoms, no space, no uniform time, no management, and no control, then maybe we've been wrong about what matters in the real world in the first place.[19]

Imagine, as John Lennon invites us, a world in which we take ourselves, and our rigid constructions of religion, nation, and work, with a grain of salt (even salt water). Imagine a world to which we hold on loosely; where what matters is satisfaction in community, work with an element of play, and celebrating life in each moment. Both the hacker and the porpoise inhabit this kind of world.[20]

> An authentic life is the most personal form of worship. Everyday life has become my prayer.
>
> *Sarah Ban Breathnach*

The value system and resulting lifestyle of hacker culture might be called holographic or holistic. Instead of a compartmentalized life of many parts, each sectioned off from the others, the wholeness of life is evident in each part, process, and relationship.

Porpoises, dolphins, and whales are way ahead of the hackers in the development of a life-centered lifestyle. They have placed life at the center of every action and interaction for centuries, and even the act of finding and harvesting food is joy-filled:

> Man has devoted his intelligence to adapting [manipulating] the environment to suit his needs. The majority has little time to play and enjoy themselves. Cetaceans, on the other hand, have themselves adapted through the process of evolution to harmonize with their environment. A dolphin [porpoise] lives in an ecosystem in which food is abundant. He has no need to construct a shelter. Unburdened by possessions, avarice must be unknown to him. With the open sea as a common heritage, cetaceans do not suffer the tensions of living in high-rise blocks of flats, the resentments of the squalor of shanty towns, or the burning hatred of suppression in the ghettos. Having no money with which one dolphin can exploit another, they have no problem with the corrupting influence of power. No dolphin has a fortune whilst another suffers the misery of poverty, and no starving dolphin swims alongside a bloated glutton. Unspurred by the greed for land, no dolphin technology is needed to develop mustard gas,

> flame throwers, defoliants, [nuclear weapons] or any of the other hideous instruments of human warfare. . . .[21]

With no destructive technology, no artifacts, no monuments or kingdoms:

> One might ask, for what purpose did the cetaceans evolve their large cerebral cortex through the past [three hundred million] years? One answer to that question might be that the dolphins have evolved in order to enjoy and revel in the pleasure of simply being alive, of being dolphins.[22]

Sweet points out in his exposition of Judeo-Christian Scripture that one of the first commandments of God, in Genesis 2:16, is to eat. The last biblical commandment of God, in Revelation 22:17, is to drink. In between is a dramatic, passionate, righteous, gritty, real, and hopeful smorgasbord of life. In his book *The Gospel According to Starbucks*, Sweet explains:

> Like the word "love" which in Greek has at least four words to describe its full range of meaning, the word "Life" also has many words in Greek: When we speak of Life here, we do not speak of mere "bios" (or biology), but "Zoë," as in flamboyant, passionate life."[23]

In the midst of it, Jesus says, "I have come that you would have *zoë* and have it more abundantly" (John 10:10).

One man who lost a loved one in the terrorist attack on the World Trade Center offers this advice to those who seek to honor those who died on September 11, 2001: "Celebrate their life by celebrating all life, including your own. Take a moment of silence to remember, but spend the rest of the day in celebration. There is no greater way to honor someone than to live, love, and laugh in their honor."[24] This is exactly what the Creator is asking you to do—live, love, and laugh in honor of the giver of life.

The invitation of Koheleth and the porpoises, of God the Creator, God the Holy Spirit, and God the Child, is not an invitation to gluttony and drunkenness, but to the joy of good food, good times, good company, and good work. What are we to do with this life we have been given? Celebrate the fact of it. Live while you are living.

"Life Is . . ."

by Mother Teresa

Life is an opportunity, benefit from it.
Life is beauty, admire it.
Life is bliss, taste it.
Life is a dream, realize it.
Life is a challenge, meet it.
Life is a duty, complete it.
Life is a game, play it.
Life is a promise, fulfill it.
Life is sorrow, overcome it.
Life is a song, sing it.
Life is a struggle, accept it.
Life is a tragedy, confront it.
Life is an adventure, dare it.
Life is luck, make it.
Life is too precious, do not destroy it.
Life is life, fight for it.

The Reflecting Pool

The Reflecting Pool

Quotes to Ponder

The happiness of the bee and the dolphin is to exist. For man it is to know that and to wonder at it.

Jacques Yves Cousteau

What we call the secret of happiness is no more a secret than our willingness to choose life.

Leo Buscaglia

To do is to be.—Descartes
To be is to do.—Sartre
Do be, do be, do.—Sinatra

Anonymous

Questions to Consider

1. Which comes first, being or doing? How so? What difference does it make? Could it work both ways?

2. How do you respond to the assertion that the purpose of life is life itself? It is not in essence either a test or a dress rehearsal for eternity.

3. What, if any, is the difference between living and existing?

Actions to Take

1. Don't just do something, sit there. Take time daily to be still.

2. Read *Original Blessing* by Matthew Fox.

3. Meditate on the idea of *imago Dei*, being created in the image of God.

Biblical Wisdom

Know that the LORD is God. It is he that made us, and we are his . . .

Psalm 100:3

A Breath Prayer

Teach me how to pray and teach me how to live. Make them one and the same that my life is a prayer and my prayer is for life.

Chapter 8

Delight

There is a river whose streams makes glad the city of God . . .
Psalm 46:4

God delights in steadfast love.
Micah 7:19

Maybe you've seen (or perhaps you have) the popular set of encouragements that have been made into jewelry, printed on T-shirts, and tagged onto e-mail signatures that reminds the bearer and invites the reader to "Live, Laugh, Love." It's simple shorthand for a profound philosophy of life. In the last chapter, we celebrated the gift of life. In this chapter, we'll celebrate receiving and giving the gifts of laughter and delight. In the next, we'll explore active love—love with hands, feet, and fins.

Fun with a Porpoise

Cetaceans spend three times more of their time playing than they do searching for food! And they play without any other goal than to play.[1] They even make searching for food into a game. Heathcote Williams states:

> Cetaceans unquestionably have big brains and the frequency with which they use them in patterns that can only be described as play suggests that they frolic with their minds as readily as they do with their bodies. This tendency towards mental playfulness may in itself been partly responsible for the enlargement of their brains.[2]

The word "porpoise" has entered American English slang as a verb: "in mountain biking, to ride responding to rather than controlling the bike."[3] It is listed as a verb in the *Merriam-Webster's Online Dictionary*, too: "To leap or plunge like a porpoise."[4]

Coined after the smaller of the two subspecies of cetaceans, dolphins "porpoise," rising out of the sea in low leaps that keep their head clear of the water so they can breathe. It also looks like a lot of fun! Spinner dolphins get their name from their habit of leaping out of the water while spinning on their long axis. Trained porpoises and dolphins can leap straight up as high as twenty feet. Although rare in large cetaceans, some rorquals, or baleen whales, like the humpback whale seen in those Pacific Life Insurance commercials, have been photographed jumping clear of the water.

We are thrilled to spot dolphins playing in the wake of our boats, bodysurfing and bowriding, to both their delight and ours. These fun-loving and delight-bearing cetaceans can even catch a good bow wave from their larger cousins, when baleen whales manage to swim fast enough to produce a wake!

Unlike the "constipated" congregants of Duffey Robbins's childhood church, cetaceans aren't shy about their PDJ (public display of joy). They invite us into a no-holds-barred, bare-chested boisterousness. The porpoise-given life invites us to live, laugh, and love—with an accent on laugh.

Remember, the participatory bits of public liturgy are called acts of worship, not sits of worship! According to Matthew Fox, "the body has been effectively banished from most white worship in the West. That is one role stationary benches play in the church—they assure no dance, no celebration of body-spirit, might break out."[5]

Laugh 'Til It Heals

Gelotology is the study of humor and laughter. Makes sense to me . . . Jell-O makes me laugh.[6] Laughter actually lowers blood sugar levels after a meal! Laughing one hundred times equals a ten-minute workout on a rowing machine or fifteen minutes on a stationary bicycle, and one good belly laugh will raise your heart rate faster than the bike.[7]

> The most wasted of all days is one without laughter.
>
> *e. e. cummings*

Belly laughing lowers blood pressure, reduces levels of stress-producing hormones, relaxes muscle tension, and boosts the immune system by enhancing the production of infection-fighting T cells and disease-fighting proteins gamma-interferon and B cells. Laughing releases endorphins, the body's natural painkillers, and produces a general feeling of well-being.[8]

Therapist Marilyn Sprague-Smith teaches her clients that laughter prevents "hardening of the attitudes,"[9] and Elva Dodd, a frequent visitor to laughter classes in Seattle, encourages participants to "laugh 'til it heals."[10]

I carry around in my heart a video clip of hope-filled play from the bombed-out city of Baghdad, Iraq. It's an image of Shane Claiborne, friend and cofounder of The Simple Way,[11] playing soccer with a group of Iraqi children as bombs explode in the background. Amid the sounds of war, you can hear the sound of laughter. I'm not suggesting that the war will simply go away if the children whistle a happy tune, but this image is a powerful testimony of the healing and sustaining power of simple pleasures in the midst of very real pain and suffering.

We are not to walk around in sackcloth and ashes, preoccupied with the struggles of life. Neither are we to seek mere creature comforts and the easy way.

> Neither the hair shirt
> Nor the soft-berth
> Will do.
> The place God calls you
> Is where your heart's deep gladness
> And the world's deep hunger meet.[12]

If not its goal, certainly a sweet fruit of all human endeavors is joy. Oprah Winfrey claims, "The more you praise and celebrate your life, the more there is in life to celebrate."[13]

> The person born with a talent they are meant to use will find their greatest happiness in using it.
>
> *Johann Wolfgang von Goethe*

Do you know someone with joie de vivre, the joy of life? Someone who lights up a room when they enter, someone who is so delightful that you delight in them? Like the porpoise that jumps beside our boat or leaps over its bow, they offer their gift of delight—pure, uninhibited, and without condition or price.

Like them, do you want to delight others? *How?* you may ask. Well, how do they delight you? By taking delight in you. By taking delight in others. If you want to delight others, be overjoyed to see them. Be eager to hear about their lives and celebrate with them. Delight in their success. Know them well enough to know what brings them joy and become a source of joy for them. Experience joy yourself and do not hide your delight. Joy, like gloom, is contagious. It's one of those good viruses, passing from person to person.

Become a delight-full person. Have a ball, but don't hog the ball. Pass it. Help someone else have a ball. Pass the ball, not the buck. Get the ball rolling. Keep the ball rolling. Live life less as a ball and chain and more like a disco ball! In the face of real challenges, do what preacher Motlalepula Chubaku suggests: be more like a bouncy tennis ball—the harder you slam it down, the higher it bounces back up![14]

Make God Smile

Leonard Sweet delights in word play. He contends that most of our prayers are bidding prayers, prayers that ask for things like healing, forgiveness, peace, money, good grades, traveling mercies, and safety on ball fields and courts. All of these prayers begin or end with the phrase, "Please, God. . . ." Sweet asks, "What if the motivating passion behind our prayers and the living of our lives, day to day, moment by moment, was to please God?" He even had little cards imprinted with the simple invitation to "Please God" with the comma removed. In fact, you'll know how to better pray "Please, God . . ." if you'll get in tune with what pleases God.[15]

Koheleth, the Preacher of Ecclesiastes, encourages us to delight God by enjoying the gifts and resources that come our way and by sharing them with others. Other things that delight God include:

Creativity

Creativity is the first attribute of God demonstrated in the Bible. Creativity may be God's most fundamental power. Has God stopped creating? No! The United Church of Canada's Affirmation of Faith declares, "We believe in God, who has created and is creating still!"[16] My wife—artist, deacon, and workshop leader Gloria Hughes—encourages others to delight God by recognizing and exercising the creative spirit within.[17] In fact, you are cocreating your life each day with every thought, every experience, and every response. Some create great food. Some create great spaces for hospitality and community. Some create music, movement, or visual art. Some create with words and some with visual storytelling. But since we are all created in the image of God (Gen. 1:26), we are all endowed with creativity. Artist Catherine Kapikian affirms, "An artist is not a special kind of person. Every person is a special kind of artist."[18] Delight God by expressing yourself, being creative in your way, and in the exercise of cocreating with others.

The Creation

God delights in the world that he has made. Genesis 1:31 declares, "And God saw every thing that he had made, and, behold, it was very good" (KJV). Delight God by delighting in the natural world. It is wondrous to behold.

Faithful Stewardship of the Creation

It pleases God when we are faithful stewards of creation, enjoying it with gratitude and tending it with care. Of course, the opposite is also true. It displeases God when we are careless with creation, fouling the air and the water, depleting the land, and killing hundreds of living species to the point of extinction.

Doing New Things

The prophet Isaiah gives voice to God who declares, "Behold, I am doing a new thing; now it springs forth, do you not perceive it?" (Isa. 43:19). Delight God by being a lifelong learner. Try new foods, meet new people, have new experiences, and entertain new ideas.

A Confessional Life and a Contrite Heart

Humility and a confessional lifestyle are rare in the USAmerican psyche. We'd rather bow up than bow down when we are confronted with our sins. But God delights in a kind of trusting transparency. What awaits us is forgiveness and renewed relationships. "If we confess our sins [God] is faithful and just and will forgive us our sins and purify us from all

> What I have done I have done, because it has been play. If it had been work I shouldn't have done it. . . . The work that is really a man's own work is play and not work at all. Cursed is the man who has found some other man's work and cannot lose it. When we talk about the great workers of the world we really mean the great players of the world.
>
> *Mark Twain*

unrighteousness" (1 John 1:9 NIV). God delights in blotting out our transgressions, and forgetting our sin. Isaiah 43:25 declares, "I, I am He who blots out your transgressions for my own sake, and I will not remember your sins." Psalm 51:17 assures us that the sacrifices which delight God are a contrite heart and a humble spirit.

Extending to Others What We Have Received

It pleases God when we do for others what God has done for us, namely, show mercy, extend grace, and forgive sins. Remember Sweet's Titanium Rule: "Do unto others as Christ has done unto you."[19] We seldom hear the ending of what is called The Lord's Prayer as recorded in Matthew 6:9–15. Right after "and deliver us from evil," Jesus goes on to say, "For if you forgive men their trespasses, your heavenly Father also will forgive you; but if you do not forgive men their trespasses, neither will your Father forgive your trespasses." To quote Shakespeare's Hamlet, "Aye, there's the rub."[20]

Siding With the Poor and Marginalized

Blessed is the one who considers the poor and the helpless, according to Psalm 41:1. God delights in those who consider the poor, who side with the poor, and who respond in concrete ways to the poor. Delight God and cultivate relationships with the poor and those who live on the edges of your community. Listen to them. Experience life through their eyes. And respond with advocacy and hands-on mercy. Jesus reminds us, if you clothe, feed, welcome, shelter, visit, or extend respite and refreshment to any one of these, "you did it to me" (Matt. 25:40).

Servanthood

Jesus measures success by the standard of service. In fact, he says (and demonstrates), "The greatest among you will be your servant" (Matt. 23:11 NIV). Delight God by being a servant in all things. Cultivate the lifestyle of servanthood.

Justice

Even more than worship and praise, justice is pleasing to God. The prophet Amos says it this way:

> I hate, I despise your feasts, and I take no delight in your solemn assemblies. Even though you offer me your burnt offerings and cereal offerings, I will not accept them; and the peace offerings of your fatted beasts I will not look upon. Take away from me the noise of your songs; to the melody of your harps I will not listen. But let justice roll down like waters, and righteousness like an ever-flowing stream. (Amos 5:21–24 RSV)

> Joy lies in the fight, in the attempt, in the suffering involved, not in the victory itself.
>
> *Mohandas Gandhi*

Knowing what delights God and doing it will make us more delight-full people. The God of Abraham, Isaac, Jacob, and Jesus speaks through the prophet Micah to reveal what delights God, "to do justice, and to love kindness, and to walk humbly" (Micah 6:8). It can be done. We can do it together.

> Joy is not in things; it is in us.
>
> *Richard Wagner*

No Excuse

Most of us have said it at one time or another as we cry out for understanding, forgiveness, or a little slack, "I'm only human!" *Only*? We are fearfully and wonderfully made, just a little lower than the angels, incredibly strong when the going gets tough, creative, ingenious, and indomitable. We are created in the image of God and pronounced "very good" by our maker (Gen. 1:31). The problem is not that we are only human, but that we settle for being less than human. The human creature pleased God at its creation.

The human creature displeases God when it is not human (humane, noble, just, creative, and compassionate). We are not just human, we are called to be *just* humans.

In the little gem of a book, *Ten Fun Things to Do Before You Die*, Sister Karol Jackowski writes, "What I celebrate is how divine all of life's experiences can be. You know, we celebrate Christmas at the time of the Winter Solstice, the darkest night of the year, and we celebrate the birth of Christ to signify the light in the darkness."[21] Joy is finding the light in the darkness. The playful nun says that "fun" is her favorite "F" word!

Number one on her top-ten list is "Have more fun than anyone else." In other words, in any situation or setting, be satisfied that you are having all the fun there is to have. Make it happen. Find the fun. Create the fun. Be the fun. Her other pre-mortem suggestions include:

- Find the fun in unexpected moments and circumstances.
- Find fun people.
- Don't think about yourself around other people—focus on them.
- Be a fun person.

And finally, "if it looks like fun and it doesn't break the Ten Commandments, *do it*!"[22] The porpoise and the dolphin invite us to find the joy in living, and, as delight-full creatures, delight others.

The Reflecting Pool

Quotes to Ponder

Enjoy life. This is not a dress rehearsal.

Anonymous

Life is not measured by the number of breaths we take, but by the moments that take our breath away.

Anonymous

Questions to Consider

1. What brings you the greatest amount of satisfaction? Why?

2. Do you spend most of your time doing it? Why not?

3. Is there a difference between happiness and joy?

Actions to Take

1. Start keeping a "Joy Journal."[23] For six weeks, record some source of joy from your day, even if it is one small moment—a taste, a sight, a smell, a touch, a sound, an encounter.

2. Make a pact with a friend or family member to share the funny things that happen (or the funny things you saw, heard, or read) during each day.

3. Spend most of your time doing what brings you the greatest amount of joy.

Biblical Wisdom

You turned my wailing into dancing; you removed my sackcloth and clothed me with joy.

Psalm 30:11 (NIV)

Go and enjoy choice food and sweet drinks, and send some to those who have nothing prepared. . . . Do not grieve, for the joy of the Lord is your strength.

Nehemiah 8:10 (NIV)

Love does not delight in evil but rejoices with the truth.

1 Corinthians 13:6 (NIV)

Here is my servant, whom I uphold, my chosen one in whom I delight; I will put my Spirit on him and he will bring justice to the nations.

Isaiah 42:1 (NIV)

A Breath Prayer

I celebrate the joy of life. May I delight in others and be delight-full today.

Chapter 9

Save

Greater love has no man than this,
that a man lay down his life for his friends.
John 15:13

. . . am I my brother's keeper?
Genesis 4:9

Let's return to the introductory image of dolphins as lifeguards. The astonishing role of the dolphin as a lifeguard and life giver to humans who find themselves literally over their heads. . . . It's an image that both amazes and inspires us, but to what degree?

There are lots of humans who are over their heads, under their debts, without food and clean water, and beside themselves with fear and grief. We'd like to think that if we knew about those needs of another person, we would do all we could to help them. But let's face it, most of the time we are not aware of the needs of those around us and around our shared world. And when we are made aware through television reports, magazine articles, and e-mail

> . . . As far back as 1933 I observed a school of dolphins (their schools increase as ours decline) romping, as we carelessly call it, alongside a cruise ship in the South Atlantic, and something told me that here was a creature, all gaiety, charm, and intelligence, that might one day come out of the boundless deep and show us how a world can be run by creatures dedicated not to the destruction of their species, but to its preservation.
>
> *James Thurber*
>
> *quoted in* Whale Nation

forwards, we leave it to others to step up and help out. We like to think that if we have a heartfelt response to a special report on global poverty, or the plight of refugees, or to a five-minute plea from a world relief organization filled with pictures of starving children in the middle of late-night television, that we really care and that we have, in fact, done something.

What's truly astonishing is that we have the global resources to alleviate most of the basic human crises on the planet, and we choose to squander our innate creativity and ingenuity on entertaining and killing ourselves. The question remains: will we overcome our preoccupation with ourselves and begin to lay down our lives for our neighbors? Will we be creative and cooperative in the face of crushing human suffering and global environmental challenges?

Have a Heart

Cetaceans have a four-chambered heart, just like all mammals. But whales, dolphins, and porpoises have a unique "marvelous network"[1] of reservoirs for collecting extra supplies of oxygenated blood that keeps their core warm and alive in extreme temperatures and changes in environment. This enables adequate blood flow to the muscles used for diving and swimming while using the oxygen stored in the reservoirs to maintain the heart and brain (the two organs that depend on a constant supply of oxygen to survive).

You can divide the sins of man into hot sins and cold sins. Hot sins are sins of passion and hot-blooded appetites such as adultery, substance abuse, violence, gluttony, and revenge. Cold sins are just as deadly, but much less sexy and entertaining. These include indifference to poverty, couldn't-care-less-ness toward hunger, smugness toward the homeless, and a cold shoulder to the stranger in our midst.

We need to learn how to tend the core of our being—our heart of compassion and empathy and our mind focused on creative, interdependent living—so that we do not become cold: cold-blooded, cold-hearted, and prone to cold sins. We need to develop and maintain a marvelous network of global awareness, prayer, and friendships with communities different from our own to keep a constant flow of life running through our lives.

Cetaceans are constantly monitoring the world around them. What's going on around you? How aware are you? One gauge of maturity is the realization that there are other people in the room (or in the world) besides you, your tribe, and your country. Another is the development of a deep sense of "otherness" that moves you to actively respond to the needs of others at a real cost to yourself.

No Greater Love

Dolphins are extremely protective of their young (as are most animals, including the human kind). They have been observed following the captors of their calves, putting themselves at great risk, allowing themselves to be beaten, speared, and killed in their efforts to save their offspring. No greater love has a creature than this, to lay down its life for another. Can we enlarge our definition of "our children" to include "all children?"

Many cetaceans exhibit epimeletic behavior, in which healthy animals take care of another animal that has become temporarily incapacitated. You may be familiar with this phenomenon in flocks of geese in their migratory journeys. A healthy member of the flock will descend to the ground with a wounded or disabled member until the injured goose is again able to fly. Its caregiver is there for protection and support. The caregiver stops its own progress to journey with one in need. This is a striking example of laying down one's life for another. Laying down your life doesn't always mean that you have to literally die. It can mean that you lay down your preferences, your resources, your time, or your comfort for the sake of another person or creature.

> When one's own problems are unsolvable and all best efforts frustrated, it is lifesaving to listen to other people's problems.
>
> *Suzanne Massie*

In our interdependence, we are challenged and privileged to be part of someone else's salvation. Yes, salvation. Stay with me here. "Salvation" is a big word. It cannot be confined to concern over someone else's eternal soul, especially when their bodies and minds may be so impoverished in this present life. It includes the salvation (restoration) of the whole person—not just a soul, but as a living being in the here and now.

The word implies salvation from one thing to another. If you are drowning, you hope to be saved from the water into a boat. There is salvation from despair to hope, from ignorance to awareness, and from loneliness to companionship. People long to be saved from poverty by being empowered to earn a living wage or to own a piece of fertile land. Until we believe that we are called to become at least part of the answer to another person's prayers, literally part of their salvation, we will leave them drowning in their circumstances.

According to Abraham Maslow's famous hierarchy of needs, before a person or a people can gather the energy and the vision to develop a healthy self-image, dream dreams, and set goals beyond simple survival, their basic bodily needs have to be met. These include food, water, and a fairly constant body temperature that comes from shelter and clothing. Second is a sense of safety and security. [2]

"Rescue the Perishing, Care for the Dying"[3]

Cetaceans respond to fear by either fleeing, often en masse, or by bunching up and milling (literally) around. Fishermen can trigger the flight response as a technique for herding them into nets. Once trapped, schools of dolphin may begin to move in a circular pattern or as an undulating mass, and at the height of this reaction they stop swimming, sink, and die.[4]

Politicians have used the same technique to snare people into webs of misguided policy and fiscal irresponsibility. Fear is not a very reasonable emotion. Although authentic fear can be useful in keeping a person from danger, manufactured fear can cause a person to do unreasonable things and to believe anyone who is skilled at selling a solution. Remember, a sense of safety and security is necessary for a person or a people to be able to see clearly, dream of a preferred future, set reasonable goals, and make plans for getting there.

> If you think somebody cares about you and believes your life is worth saving, how can you give up?
>
> *Geraldine Ferraro*

> Saving a life overrides territories.
>
> *Ovadia Yosef*

Accidental entanglement (bycatch) in fishing nets is the main threat to porpoises today.[5] Millions of innocent men, women, and children are trapped and killed as bycatch in civil wars, regional conflicts, and international police actions in support of political experiments. The working poor are caught as bycatch in business practices and governmental policies that maximize profits and marginalize people. Hundreds of extinct or endangered species are sacrificed as bycatch in the march of consumer civilization and the abuse of our shared natural resources of air, soil, and water.

No doubt you've seen pictures or news reports of beached whales. We can't always tell why a single whale or a whole pod of porpoises strand themselves, but there appear to be a number of reasons.[6]

The Leader of a Group Goes Astray

Been there, done that. Sometimes we follow our leaders into deadly and sinfully wasteful waters because following the leader is often valued above loss of life and increasing national debt in our American culture. You can take a lot of flack for refusing to follow the designated leader and following the lead of your own heart and mind. Sometimes the emperor really is naked.

Epidemic Disease

It's no wonder that whole populations can "stop swimming, sink, and die" when faced with epidemics of diseases that we thought we had eradicated, simply for the lack of clean water. We have the means to eradicate most epidemic disease if we have the will. A single dose of the measles vaccine costs a mere fifteen cents. For the lack of enough fifteen-cent measles shots, a hundred thousand children die each year in India. The Measles

Institute, a coalition of charities, world health organizations, and the United Nations, estimates that it can reduce global measles by 90 percent with this inexpensive vaccine at a total cost of $479 million. Compare that to the $9 billion per month, about $300 million per day, that it costs to sustain the war in Iraq. Less than two days of war expenses could save ninety thousand children.[7]

Getting Lost in Pursuit of Prey

Here again, it's easy to get lost when you are motivated by revenge. Our government can justify spending $4 trillion for a seven-year war and vetoing the continuation of a state-supported, bipartisan, highly successful program to insure millions of children at a cost of $6.2 billion spread over five years, just $1.2 billion over the previous five years of the program,[8] if you are blinded by revenge and pride.

Following Historic Migratory Routes

We can learn from our past and honor the ways of our forebearers, and even share goals as a people that have a long history without necessarily doing everything the way it has always been done. Remember the seven last words of a dying organization: we've never done it that way before. Albert Einstein cautions, "We can't solve problems by using the same kind of thinking we used when we created them."[9]

Magnetic Anomalies that Lead the School Astray

When we begin to stray away from our true north, we can lose our way. Pride and arrogance will make it difficult to make a course correction. We are too good at rationalization, perhaps our greatest sin: learning to call a good thing bad and a bad thing good.

Fright Reaction to Predators

We simply must break our susceptibility to the use of fear as the primary tool in the arsenal of our political leaders. Threats to our personal and corporate safety may be very real. However, we need to get the facts and make reasonable decisions in the face of such dangers.

Failure of Echolocation in Shallow Water

We can avoid false echoes that can lead us astray in the shallows of pop culture and incessant sales pitches by seeking deeper waters and by developing reliable depth finders. Critical thinking and spiritual disciplines of prayer, meditation, and discernment will help fine-tune our "BS meters" to help us separate truth from half-truth and the truly important from the merely urgent. Practice asking yourself and your trusted companions, "What's going on beneath the surface?" as you listen to political speeches, watch movies, and enjoy humor. What's going on beneath the surface of your outward appearance? How is it with your soul?

Suicide

If delightful dolphins and wizened whales can commit suicide, it's understandable that most people entertain the thought at least once in their life! Although we should not own the guilt of another person's suicide by insisting to ourselves that we could have stopped them, we can be more observant, more sensitive, and more proactive in our care for one another. If a person trusts you enough to talk about it, they'll trust you enough to let you encourage them to seek help. Community and companionship can help depressed and suicidal people recognize their present location and reset their compass to a more hopeful true north.

Can we, who have our basic needs met and more, be so callous that we would at best ignore and at worst blame those who cannot meet basic needs? Apparently we can. Maybe we need a change of perspective. One giant step toward effective interdependence is the recognition that we are, in fact, our brothers' and sisters' keeper. It's time to take up the mission of Christ proclaimed in Luke 4:18, echoing the prophetic voice of Isaiah 61:1. We are called to bring release to the captives, sight to the blind, and healing to the sick. We are called to join the continuing incarnation of the Christ who challenges us to feed, clothe, shelter, protect, and respect those considered the least important, the disposable, the collateral damage of war, disease, and famine.

A Seamless Garment

In 1983, Joseph Cardinal Bernardin, former archbishop of Chicago, began what would become his greatest work in his call for "a consistent ethic of life."[10] This consistently pro-life ethic recognizes the sanctity of human life and defends it against all technologies, public policies, and attitudes that would diminish its sacred worth. For such an ethic to be truly consistent, forming what has been called "a seamless garment" of sacred stewardship, it has to be extended to include our relationship with all living things, including the living organism called the earth. Abortion, war, hunger, human rights, euthanasia, capital punishment, living wages, housing, the environment, hate crimes and bigotry, stewardship of the earth and all living things—and all of the spiritual, social, and economic ramifications of each decision—are a part of this seamless garment.

A consistent ethic of life connects all of these things, but we usually treat them as separate issues. If saving unborn children is "pro-life" then we need to prepare for its support, development, health, safety, and maturity as well. Under a consistent pro-life ethic we would consider the ramifications of saving that life. Who

> The opposite of love is not hate, it's indifference.
> The opposite of art is not ugliness, it's indifference.
> The opposite of faith is not heresy, it's indifference.
> And the opposite of life is not death, it's indifference.
>
> *Elie Wiesel*

will raise the child, feed the child, clothe the child, and educate the child? If we care about its life, we will care that it can find a job with a living wage. We will care that it has access to affordable health care throughout its life. We will care that it retains its dignity in old age. An antiabortion ethic that does not provide for the ramifications of life after birth is not a pro-life ethic.

Jim Wallis, editor of *Sojourners* magazine and outspoken prophet, writes:

> My message to both parties—to both liberals and conservatives—is that protecting life is indeed a seamless garment. Protecting unborn life is important. Opposing unjust wars that take human life is important. And supporting anti-poverty programs that provide adequate support for mothers and children in poverty is important. Neither party gets it right; each has perhaps half of the answer. My message and my challenge are to bring them together.[11]

There is a profound line in a telling scene, among many, in the movie *Pay It Forward*, in which a recovering drug addict reaches out to a woman who is about to commit suicide by jumping off of a bridge. In one simple invitation, the addict reveals the deep truth of our interdependence: "Save my life . . . let me help you."[12] It's about time we invested in each other's salvation.

> What do we live for, if it is not to make life less difficult for each other?
>
> *George Eliot*

Quotes to Ponder

Only when we are no longer afraid do we begin to live.

Dorothy Thompson

You begin saving the world by saving one person at a time; all else is grandiose romanticism or politics.

Charles Bukowski

An individual has not started living until he can rise above the narrow confines of his individualistic concerns to the broader concerns of all humanity.

Martin Luther King Jr.

Questions to Consider

1. Who is your neighbor? What, if any, is your obligation to him, her, or them?

2. In what specific ways would a consistent pro-life ethic change your thinking, attitudes, or actions?

3. How might you contribute to the salvation of another person? Be specific . . . salvation from what to what?

4. From what do you need saving?

Actions to Take

1. Investigate ways you can add your small contribution to a life-saving, life-enhancing cause like hunger relief, clean water wells, abuses of human rights, and environmental issues. Make that contribution.

2. Watch a political speech or a series of paid advertisements (commercials) with a friend and ask each other what is really being said and sold.

3. Echoing the suggestion of retired general Bernard Loeffke, take a lifesaving course that includes CPR. Become a healer.

Biblical Wisdom

So faith by itself, if it has no works, is dead.

James 2:17

A Breath Prayer

Help me be the answer to someone else's prayer today.

Forward

Being Human

Beyond the Porpoise-Given Life

But the hour is coming, and now is . . .
John 4:23

Thy kingdom come. Thy will be done,
On earth as it is in heaven.
Matthew 6:10

Then God said, "Let us make man in our image . . .
Genesis 1:26 (NIV)

The Judeo-Christian tradition celebrates the creation of human beings in the image of God, the *imago Dei*. Of course, this is not to say that God is created in our image (as some suggest)! God does not look like a man or a woman, or any other creature. God is not a person or a thing. God is. Remember, God goes by the name "I AM" (Exod. 3:14). So, being created in God's image must have more to do with our essence, our best nature, and those qualities that would reflect the uniqueness of being human. We looked at some of the clues to God's essential nature and our best selves with the discussion of delight in chapter 8. I like to think that reflective consciousness, a sense of justice, and an

appreciation of beauty, love, and self-sacrifice are all a part of the *imago Dei*.

I began this book's introduction by listing three popular images of cetaceans that persist in the imagination of mankind. Let me suggest three things, among many more, that appear to move us beyond the porpoise-given life to a life of being human on purpose:

- A sense of time
- A capacity for creativity
- A free will

Does Anybody Really Know What Time It Is?

There's an old joke passed around by psychiatrists. A guy comes in for therapy, complaining that he thinks he is a tepee, only to change his mind, declaring that he is a yurt. Back and forth he continues his indecisive rant. Finally the doctor offers this diagnosis: "My good man, you are two tents."

It appears that the animal kingdom lives in the present tense with little, if any, thought toward the future, relying on instinct and the blueprints of their DNA to stimulate and guide them in a kind of "just in time" lifestyle. We can learn a lot about living in the moment that is liberating and healing from our highly evolved friends in the sea.

Jesus reminds that we will add nothing to our life by worrying about the future (Luke 12:22–31). And yet, we have this capacity to see visions and set goals that draw us forward.

"Futurepresent" is a hybrid adjective descriptive of a perspective that sees the future emerging in the present, signifying a proactive stance toward discerning and cocreating a preferred future.[1] This is the natural perspective of God's people:

> In what may be the most important study on eschatology in the twentieth century, Jürgen Moltmann argues that God's middle name is "Future," that God speaks to us from the future and calls to follow and move forward from the future. We worship

> a God for whom, the future is [God's] essential nature. In the words of Pauline scholar Paul J. Achtemeier, commenting on Romans 13:11–14, Christians are "creatures of the future" not the past. To it they are to look and by it they are to act."[2]

It is the perspective of Jesus when he says, "The hour is coming [future], and now is [present]" (John 4:23). This "futurepresent" orientation stands in stark contrast to a "pastpresent" orientation common to many institutions.

> It is better to create than to be learned, creating is the true essence of life.
>
> *Barthold Georg Neibuhr*

"Pastpresent" describes a perspective focused on the preservation and maintenance of past traditions, processes and systems. It asks, "How can we keep our present like our past?" It longs for the "good old days" and assumes the future will be pretty much like the past and present, with a few grudging adjustments. The past of this perspective is not ancient past, but sixteenth-century, Reformation, Enlightenment past.[3]

It is the polar opposite of a futurepresent orientation as it seeks to ignore, or at least discredit, the signs of the times. The pastpresent perspective denies the impact of forces, trends, technological innovations, and changes external to its institutional culture. Technology provides new tools to do old things.

A "pastfuture" orientation is an extreme change-averse position. Any adaptation to change is viewed as anathema, betrayal, and loss.[4]

A "presentfuture" perspective is focused on short-term, evolutionary change, forced to adapt to realities coming from the world at large and surrounding cultures that begin to affect institutional culture. It asks, "What from our present will survive into the future?" It adapts out of necessity and belated expediency. Presentfuture mimics futurepresent but has more in common with the pastpresent mindset. It may appear to be progressive from within, but it only allows institutions to be less behind than they might be. It never leads. It only responds.

Jesus is present in all three tenses—past, present, and future (Heb. 13:8). Part of being created in the *imago Dei* is our capacity to be present in all three tenses with the aid of both memory and imagination. One indication that we are not gods is our perspective of time as a linear phenomenon. Events, including our lives, have a beginning, middle, and end. It's this sense of ending that motivates us to live with purpose.

And yet, we can conceive of being that is everywhere and everywhen. We intuit that we are part of what Paul Tillich calls "the eternal now."[5] We care about our legacy and we hope to leave the world better for our being here. A futurepresent orientation is one of hope. It is hopeful because we can learn from our past and imagine a preferred future. This gives us meaning in the present.

Creativity Is the New Omnipotence

The most essential attribute of God is being. And the most operative power of God is creativity. In Isaiah 43:19, God declares through the prophet, "Behold, I am doing a new thing; now it springs forth, do you not perceive it?"

We need to rethink our understanding and experience of God's power. Creativity is more powerful than destruction or brute force. We tend to think of God's power in terms of control.

> Another word for creativity is courage.
>
> *George Prince*

When we say, "God is in control," our language betrays our need for someone to take care of us, to rescue us, and to ultimately "win" in the cosmic battle of Good versus Evil.

It's time to reimagine God's true power as creativity. God makes something, even everything, out of nothing. Now *that's* creativity! We need to shift our emphasis from God who *over*-powers to God who *em*powers, and from God's *de*structive might to God's *con*structive movement. This creative power, the ability to make possibility out of impossibility, will move us from fear to faith. And it just may help us be transformed from a people who see violence and war a necessary evil or even as just and godly, to a people who persist in finding creative ways to solve our problems and settle our differences.

Co-Creation in Community

We are created for community. In both biblical creation stories (Gen. 1:1–2:3 and 2:4–23), two people are created for obvious reasons of procreation and because "it is not good for the man to be alone" (Gen. 2:18 NIV). Some affirm that the desire for companionship prompts the Creator to share essential being and create living creatures. The theological construct of the Trinity (Creator, Catalyst, and Avatar) is in itself a community of the Holy—creative, active, sustaining, and present.

Quantum science reveals that everything exists in relation to something else or it doesn't exist. Everything that can be known can only be known in relationship. Spirit "matters."[6] It materializes. It is apprehended and perceived in the midst of relationship.

And here's the real mindbender: reality is determined in relationship with observers. If God's essential nature is being, and God's most operative power is creativity, God's most astounding gift is participation—participation in creative being. This is the heart of the *imago Dei*, conscious cocreation with God in community.

Cocreativity with the Holy includes all kinds of media: paint, clay, stone, wood, dance, food, light, and virtually everything that

> Creativity can solve almost any problem. The creative act, the defeat of habit by originality overcomes everything.
>
> *George Lois*

is made from the raw materials of the universe. We also create with ideas. Social structures, governments, faith communities, institutions, and movements are also artifacts of cocreation. We also cocreate our preferred visions of the future.

"The best way to predict the future is to invent it."[7] Alan Kay, the conceiver of the laptop computer and the architect of the system that supports windows processes, called GUI (graphical user interface), reportedly blurted out his famous bit of prophetic wisdom at a staff meeting early in the history of his Palo Alto Research Center (PARC) in what has become California's Silicon Valley. Permit me to alter Kay's prophetic phraseology a bit and affirm that the best way to predict the future is to cocreate it with God, in faithful community.

Free Will-y

The operating system for this incredible gift of participation with God depends on the kernel of free will at its core. For participation to be authentic, it has to be chosen. Like the dolphin, it cannot be coerced. The system must be completely "open source." Open source software is offered without cost to the users and its program is not proprietary property, protected and copyrighted by the designer. It defies the old logic of proprietary knowledge as power. Instead, open source philosophy espouses shared knowledge as empowerment. Jesus proclaims, "All that I have heard from the Father I have made known to you" (John 15:15 RSV). Nothing is withheld. We have eaten the fruit of the tree of the knowledge of good and evil. We continually choose, and in our choosing (or failure to choose) participate in the unfolding of our future.

Being human is not the problem. Settling for being less than human, less than what we are created to be, is the problem. Forget "I'm only human." Embrace "I am human, created in the *imago Dei*. I want to be more fully, wholly, human." Remember, "And God looked at what God and made and said, "This is very good" (Gen. 1:31 paraphrased). Not just okay. Not just adequate. Not just fixable. Excellent. "A little lower than angels" (Ps. 8:5 KJV).

The brain, in itself, is not that smart. It is an excellent pattern recognizer, experience organizer, and repeater of those organized patterns. Perception is, or becomes, our reality. Over time, we hear what we expect to hear, see what we expect to see, feel what we expect to feel. We process new experiences in relationship to old ones (everything, even reality, exists because of relationships). It makes sense that we repeat behaviors that work for us and confirm our version of reality. But we repeat patterns of behavior, even if they yield less than desirable results, because those patterns are familiar.

You've heard the old joke, "Doctor, it hurts when I do this," to which the doctor replies, "Well, stop doing it." That's easier said than done, but it can be done! By choice and the exercise of will, with the help of others in our community of support and accountability, and partnership with the Holy Spirit, we can retrain the brain and establish new patterns, creating a new reality for ourselves. This, too, is an act of conscious creativity, a participation in being and will, "will-being."

> Creativity involves breaking out of established patterns in order to look at things in a different way.
>
> *Edward de Bono*

As we cocreate our preferred future, we may move beyond cooperating out of enlightened self-interest, although that's a start. We may embrace interdependence by choice, but most assuredly by necessity. Endowed with conscious being, the capacity for reflective purposeful creation, and freedom of will, let us choose to live in the futurepresent and be about the cocreation of our preferred future.

Yes, we can learn a lot about going with the flow and delighting in this life from our cetacean kin. They can teach us much about listening and navigating in the dark. They certainly know how to play! We tend to work at our play, making it about winning and losing. Porpoises do what comes naturally. And what comes naturally to them appears to be "good." They have eaten from the plankton of the knowledge of good and evil. Ever hear of a "fallen" dolphin?

We are human. We have the wonder-full, awe-full gift of conscious will. I love dolphins. I like to imagine that they like me and take delight in delighting. I'll receive the wisdom of the cetacean nation with humble gratitude. But, instead of trying to be like porpoises, I choose to be more human, on purpose and with passion.

> When we understand that man is the only animal who must create meaning, who must open a wedge into neutral nature, we already understand the essence of love. Love is the problem of an animal who must find life, who must create a dialogue with nature in order to experience his own being.
>
> *Ernest Becker*

Quotes to Ponder

Sometimes you gotta create what you want to be part of.

Geri Weitzman

Imagination is the beginning of creation. You imagine what you desire, you will what you imagine, and at last you create what you will.

George Bernard Shaw

Questions to Consider

1. What might change in your life and in the life of faith communities if we embraced the idea that God's power is defined not so much as control but as creativity?

2. Is free will real? Is it a blessing or a curse?

3. What does the idea of the *imago Dei* mean to you?

4. What other attributes and characteristics would you add to a sense of time, a capacity for creativity (including reflective imagination), and a free will to the menu of things that come with being human, made in the image of God?

Actions to Take

1. Write down a description or draw a picture of your preferred future. Share it with your circle of friends and invite them to share theirs. Run it through the filters of your faith tradition or spiritual senses. What can you do immediately that will move you toward that vision? Do it now.

2. In what patterns or loops of behavior and attitude do you feel stuck in (relationships, addictions, compulsions, self-talk, negative feedback loops) from your past? Get some help in reflecting on specific ways to break the cycles, practice new patterns, and retrain your brain (and heart).

Biblical Wisdom

What is man that you are mindful of him, the son of man that you care for him? You made him a little lower than the heavenly beings. . . .

Psalm 8:4–5 (NIV)

A Breath Prayer

I am alive. All will be well.

Notes

Preface

1. As recorded on my CD, *Hourglass*, available at www.chrisbhughes.net.

2. Rick Warren, *The Purpose-Driven Life: What on Earth Am I Here For?* (Grand Rapids, MI: Zondervan, 2002).

3. Robert M. Price, *The Reason-Driven Life: What Am I Here on Earth For?* (Amherst, MA: Prometheus Books, 2006).

4. William S. Dahl, *The Porpoise Diving Life* (Redmond, OR: William S. Dahl, 2005) http://theporpoisediving life.com. (accessed August 8, 2007).

5. Heathcote Williams, *Whale Nation* (New York: Harmony Books, 1988).

6. Scott Taylor, *Souls in the Sea: Dolphins, Whales, and Human Destiny* (Berkeley, CA: Frog, Ltd., 2003).

7. Mark Carwardine (ed.), *Whales, Dolphins, and Porpoises*, 2nd ed. (London: Dorling Kindersley, 1999). Richard C. Connor and Dawn Micklethwaite Peterson, *The Lives of Whales and Dolphins: From the American Museum of Natural History* (New York: Owl Books, 1996). Sir Richard Harrison and Dr. M. M. Bryden (eds.), *Whales, Dolphins, and Porpoises* (New York and Oxford: Facts on File, 1988). Maurizio Würtz and Nadia Repetto, *Whales and Dolphins: A Guide to the Biology and Behavior of Cetaceans* (Berkeley, CA: Thunder Bay Press, 1998).

8. For example, Sallie McFague, *Models of God*, (Philadelphia: Fortress Press, 1987). McFague offers some rich metaphors for God that take into account contemporary science and serve as alternatives to models based on military forces and royal kingdoms.

9. Leonard Sweet's homepage can be viewed at http://www.leonard sweet.com.

10. Leonard Sweet, *The Gospel According to Starbucks: Living with a Grande Passion* (Colorado Springs: Waterbrook Press, 2007).

Introduction

1. "Dolphin Saves Boy's Life: Boy pushed back to his boat after fall." Scotland: *Daily Record*, August 30, 2000. http://www.eurocbc.org/ page158.html (accessed August 2, 2007).

2. Thanks to Jim Wright, via Ed Kilbourne, for the clipping of this article from the *Midland Reporter/Telegram*, July 18, 1982.

3. Williams, *Whale Nation*, 175.

4. Ibid.

5. Mark Himmonds, *Whales and Dolphins of the World* (London: New Holland Publishers, 2007). Quoted in "Dolphins in Culture," http://www .yod2007.org/en/world_of_dolphins/ Dolphins_in_Culture/index.html. n.p.

6. Jointly sponsored by the Theological School of Drew University, The United Methodist Foundation for Evangelism, and the Southeastern Jurisdiction of the United Methodist Church, February 25–28, 2002.

7. Leonard Sweet, video transcript, *Postmodern Reformation: Navigating the New World* (Lake Junaluska, NC: Junaluska Resources, 2003), produced by Chris Hughes.

8. Philip Zaleski, "The Evil that Men Do," *Parabola,* Winter 1999, http://www .seriousseekers.com/News%20and%20 Articles/article_zaleski_evilmendo.htm (accessed May 6, 2008).

9. Ibid.

10. Ernest Hemingway, *The Old Man and the Sea* (New York: Simon & Schuster, Inc., 1995), 48.

Chapter 1: An Invitation

1. The International Bottled Water Association reports $10 billion in U.S. sales of bottled water in 2006. It projects that figure to grow to $45 billion by 2012. http://www.royalspringswater.com/sector_us.html (accessed August 19, 2007).

2. The butterfly effect is described by a theoretical physicist at Cal Tech online at http://www.cmp.caltech.edu/~mcc/chaos_new/Lor_enz.html (accessed February 19, 2008).

3. Joe Temple, "Eagle", Lesson Five in the series *Birds of the Bible Study.* (Abilene, TX: Living Bible Studies). http://www.livingbiblestudies.org/study/JT60/005.html (accessed August 25, 2007).

4. Carl Sagan, *The Cosmic Connection* (New York: Doubleday, 1973), quoted in Williams, *Whale Nation,* 179.

Chapter 2: Cetacean Citings

1. Keith Howell, *Consciousness of Whales*, quoted in *Souls in the Sea* by Taylor, xxvi.

2. San Diego Natural History Museum, Fossil Mysteries exhibit summary, http://www.sdnhm.org/exhibits/mystery/fg_timeline.html (accessed September 2, 2007). David Pilbeam, *Ascent of Man: Introduction to Human Evolution* (London: Thames & Hudson, 1972), n.p.

3. Williams, *Whale Nation*, 8.

4. "Cetacean (Order Cetacea)" inside Britannica, http://www.newsletters.britannica.com/articles/jan04/cetacean.html (accessed April 11, 2008).

5. Ibid.

6. Jeremy Bentham, *The Principles of Morals and Legislation* (Amherst, MA: Prometheus Books, 1988).

7. John Cunningham Lilly, *The Mind of the Dolphin: A Nonhuman Intelligence* (New York: Doubleday, 1967) quoted in Taylor, *Souls in the Sea,* xxv.

8. Sagan, *The Cosmic Connection* (New York: Doubleday, 1973), quoted in Williams, *Whale Nation,* 179.

9. Williams, *Whale Nation,* 179.

10. Lyall Watson and Tom Ritchie, *Whales of the World* (London: Hutchinson, 1981) quoted in Williams, *Whale Nation*, 135.

11. Seema Kumar, Discovery Channel Online News, http://www.discovery.com, as cited in Taylor, *Souls in the Sea*, 298.

12. Douglas Adams, *The Hitchhiker's Guide to the Galaxy* (New York: Del Rey Books, 1995). Quoted in Taylor, *Souls in the Sea*, xxxi. (See note #6 in Preface).

13. From "The Apostles' Creed" in *The United Methodist Hymnal* (Nashville,TN: The United Methodist Publishing House, 1989), 881.

14. Taylor, *Souls in the Sea*, xxxvii.

15. Don White, "Mystery of the Silver Rings" (Kailua, HI: Earthtrust). http://www.earthtrust.org/delphis.html (accessed August 5, 2007).

16. The same as the effect called "lift" created when air passes over an airplane wing so there is less pressure on the top than on the bottom.

17. White, "Mystery of the Silver Rings."

18. Susan Milius, "Sponge Moms: Dolphins learn tool use from their mothers," *Science News*: 167, no. 24 (2005): 371.

19. Williams, *Whale Nation*, 30.

20. http://www.brainyquote.com/quotes/quotes/w/winstonchu161474.html (accessed April 21, 2008).

21. Merriam-Webster Online Dictionary, http://www.merriam-webster.com/dictionary/porpoised (accessed April 11, 2008).

22. "Delphi." *Encyclopedia Mythica*: from *Encyclopedia Mythica Online*. http://www.pantheon.org/articles/d/delphi.html (accessed September 16, 2007).

23. Taylor, *Souls in the Sea,* xxi.

24. Williams, *Whale Nation*, 88.

25. Horace Dobbs, *Follow a Wild Dolphin* (London: Souvenir Press, Ltd. 1977), quoted in Williams, *Whale Nation*. If you are thinking, *We could help them know their Creator*, I think that they might know God in their way and worship God with their lives. For they have not learned to separate the sacred and the secular, but live as one with the Creation, and thereby, with the Creator. We could learn something here.

Chapter 3: Hydrate

1. "Dauphin" is the title given to the eldest son of the king in France, the heir apparent. In Christian trinitarian theology, Jesus is the Son of the God of the universe, with whom we are joint heirs of all creation. Hence "Jesus, the Dauphin." The word "dauphin" comes from the same Latin root word for "dolphin," *delphis,* meaning "womb."

2. Sweet, *Postmodern Reformation,* n.p.

3. Mr. Spock, *Star Trek*, the original television series.

4. Arthur C. Clarke, *2061: Odyssey Three* (New York: Del Rey, 1987), n.p.

5. Edward de Bono, *New Thinking for the New Millennium* (London: Penguin Books, 2000), 22.

6. Ibid.

7. Ibid.

8. Ibid., 23.

9. Edward de Bono, *I Am Right You Are Wrong: From This to the New Renaissance: From Rock Logic to Water Logic* (London: Penguin Books, 1991), 290.

10. Ibid., 291.

11. Ibid.

12. Ibid., 292.

13. Chris B. Hughes, "The Water," from *Hourglass* (Misenheimer, NC: Pneumanaut Publishing, 2007) http://www.chrisbhughes.net.

Chapter 4: Listen

1. In a cinematic take on Jesus' encounter with the woman at the well described in John 4:1–26. [By the way, the movie is not about angels.]

2. "Cetacean/Order Cetacea," *Britannica Online*, http://www.britannica.com/eb/article-51583 (accessed June 25, 2007).

3. Williams, *Whale Nation*, 18.

4. Ibid., 19.

5. Matthew Fox, *WHEE! We, wee, All the Way Home* (Wilmington, NC: Consortium Books, 1976), 205–207.

6. Ibid.

7. "I Know Where I'm Going" by Barry Taylor. Alternate lyrics by Ed Kilbourne, http://www.edkilbourne.com.

8. Williams, *Whale Nation*, 25.

9. Don Henley and Stan Lynch, "Learn to Be Still," *Hell Freezes Over* (Geffen, 1994).

10. "A Service of Word and Table: Confession and Pardon," in *The United Methodist Hymnal*, (Nashville, TN: The United Methodist Publishing House, 1989), 8.

Chapter 5: Breathe

1. "Cetecean/Order Cetacea," *Britannica Online*, http://www.britannica.com/eb/article-51583 (accessed June 25, 2007).

2. http://www.noc.soton.ac.uk/gg/classroom@sea/facts/index.html (accessed April 13, 2008).

3. The College of Marine Science at the University of South Florida in St. Petersburg provides an overview of plankton on its Web site. http://www.marine.usf.edu/pjocean/packets/f97/plank—1.pdf (accessed February 28, 2008).

4. http://www.noc.soton.ac.uk/gg/classroom@sea/facts/index.html (accessed April 13, 2008).

5. Williams, *Whale Nation*, 28.

6. Georg Feuerstein, Phd; Larry Payne, PhD, *Yoga for Dummies* (Foster City, CA: IDG Books Worldwide, Inc., 1999), n.p.

7. Sarah Hansen, "Breathing Facts," from *In Light Times*, http://www.inlighttimes.com/archives/2006/06/breathingfacts.htm (accessed February 28, 2008).

8. B. A. Llewellyn, "Breathing Deeply Facts," http://brightlightmultimedia.com/BLCafe/Articles-BDFacts.htm (accessed February 28, 2008).

9. Rob Bell, "Breathe," *NOOMA* DVD, (Grand Rapids, MI: Nooma, 2006).

10. You can download the free e-book version of Brother Lawrence's *The Practice of the Presence of God* at http://www.gutenberg.org/etext/5657. Read more about him at http://www.christianitytoday.com/history/special/131christians/brotherlawrence.html.

11. Neal Douglas, quoted in Llewellyn, "Breathing Deeply Facts."

Chapter 6: Cooperate

1. Leonard Sweet, *AquaChurch: Essential Leadership Arts for Piloting Your Church in Today's Fluid Culture* (Loveland, CO: Group Publishing, Inc., 1999), 30.

2. Ibid.

3. Julie Gold, "From a Distance," Copyright © 1987 Julie Gold Music, BMI/Wing And Wheel Music, BMI.

4. "cetaceans/Order cetacea," *Britannica Online*, http://www.britannica.com/eb/article-51583.

5. Leonard Sweet, Brian McLaren, and Jerry Haselmayer, *A is for Abductive* (Grand Rapids, MI: Zondervan, 2003), 49.

6. Dallas Willard, *The Spirit of the Disciplines: Understanding How God Changes Lives* (San Francisco: HarperCollins, 1988), 30.

7. Mike Holly, quoted from the transcript of my D.Min. project video. *The NeoChurch.net Project*, May 2004, Drew University.

8. Brian Swimme and Thomas Berry, *The Universe Story: From the Primordial Flaring Forth to the Ecozoic Era, A Celebration of the Unfolding of the Cosmos* (San Francisco: HarperCollins, 1992), 243.

9. Michael Talbot, *The Holographic Universe* (New York: HarperCollins, 1991), 16–17.

10. Edwin Schlossberg, *Interactive Excellence: Defining and Developing New Standards for the 21st Century* (New York: The Ballantine Publishing Group, 1998), 25.

11. Ibid., 24.

12. Ibid., 80.

13. From a lecture at Drew University, Spring 2003. Used with permission.

14. David Weinberger, *Small Pieces Loosely Joined: A Unified Theory of the Web* (Cambridge, MA: Perseus Publishing, 2002), x.

15. Ibid., 10.

16. Sweet, et. al. *A is for Abductive*, 229.

17. Steven Johnson, *Emergence: The Connected Lives of Ants, Brains, Cities, and Software* (New York: Touchstone Books/Simon & Schuster, 2002), 14.

18. Ibid.

19. Ibid.

20. Ibid.

21. James Howell, Bible study from Meyers Park UMC, Charlotte, NC.

22. Adolph Frohn, the first dolphin trainer in the modern era, circa 1952. Quoted in Taylor, *Souls in the Sea*, xxi.

23. George Hunter, *The Celtic Way of Evangelism: How Christianity Can Reach the West . . . Again* (Nashville, TN: Abingdon Press, 2000), n.p.

24. Check out *The Mission*, one of my favorite movie treatments of cultural imperialism.

Chapter 7: Live

1. http://www.thinkexist.com/quotes/howard_thurman/

2. Alois Honek, Zdenka Martinkova, and Pavel Saska, "Post-dispersal predation of Taraxacum Officinale (dandelion) seed," *Journal of Ecology* 93, no. 2 (2005): 345–352. Folklore has it that the number of breaths you take to blow all of the seeds off a dandelion is the hour of the day. And, the dandelion flower is an effective barometer, opening and closing at sunrise and sunset.

3. Reminiscent of Jesus' parable of the sower and the seeds in Luke 8:4–15.

4. These percentages are extrapolated from several studies in the survival rates of frogs at various stages of their lifespan. For more, see R. A. Relyea and N. Mills, "Predator-induced stress makes the pesticide carbaryl more deadly to gray treefrog tadpoles." http://www.ourstolenfuture.org/NEWSCIENCE/wildlife/frogs/2001relyeaandmills.htm (accessed April 14, 2008).

5. Amphibian Health and Disease: Information and Practical Advice." http:///www.froglife.org/Disease.htm (accessed April 18, 2008).

6. Charles Wesley, "And Are We Yet Alive," from *The United Methodist Hymnal* (Nashville, TN: The United Methodist Publishing House, 1749), 553.

7. Lao Tzu, *The Tao Te Ching*. http://www.terebess.hu/english/tao/bynner.html (accessed April 22, 2008).

8. See http://www.thinkexist.com/quotes/howard_thurman (accessed April 14, 2008).

9. For a good antidote to this poisonous doctrine, look at Jesus, Ecclesiastes, and the article at http://gbgmumc.org/global_news/full_article.cfm?articleid=3258.

10. Horatio G. Spafford, 1873. This hymn was written after two major traumas in Spafford's life. The first was the great Chicago Fire of October 1871, which ruined him financially (he had been a wealthy businessman). Shortly after, while crossing the Atlantic, all four of Spafford's daughters died in a collision with another ship. The hymn lyrics can be read online at http://www.cyberhymnal.org/htm/i/t/i/itiswell.htm (accessed November 20, 2007).

11. From an 1870 letter, quoted in *Emily Dickinson: Singular Poet* by Carol Dommermuth-Costa (Minneapolis: Twenty-First Century Books/Lemer, 1998), 83.

12. "Thought for the Month" from *The St. John's Eagle* (Ithaca, NY: St. John's Episcopal Church, 2002). http://www.stjohnsithaca.org/Thoughts/BookOfKoheleth.html (accessed February 28, 2008).

13. Jeff Magnum, "In the Aeroplane Under the Sea," *In the Aeroplane Under the Sea*. (Merge Records: February 11, 2002).

14. Sweet, *The Gospel According to Starbucks*, n.p.

15. Pekka Himanen, *Hacker Ethic* (New York: Random House, 2001), 140.

16. Ibid., 140–141.

17. Ibid., 19.

18. Ibid., 141.

19. Weinberger, *Small Pieces Loosely Joined*, 174.

20. Ibid.

21. Williams, *Whale Nation*, 180.

22. Ibid.

23. Sweet, *The Gospel According to Starbucks*, 4.

24. "Celebrate Life: Workplace Remembrance." http://humanresources.about.com/od/healthsafetyandwellness/a/remember911_2.htm

Chapter 8: Delight

1. Williams, *Whale Nation*, 13.

2. Ibid., 114.

3. *The New Partridge Dictionary of*

Slang and Unconventional English (New York: Taylor & Francis, 2005), 1530.

4. *Merriam-Webster Online Dictionary*, (Springfield, MA: Merriam-Webster, 2005). http://www.merriam-webster.com/dictionary/porpoising (accessed March 6, 2008).

5. Matthew Fox, *The Coming of the Cosmic Christ* (New York: HarperCollins Publishers, 1988), 217.

6. Actually, the Greek root words *gela* and *gelo* carry very different meanings: *gela* is part of words that mean "to congeal" or "freeze,"and *gelo* is a part of words like "laugh" and "laughter." In 1993, medical researchers from the St. Jerome Hospital in Batavia performed an experiment on a bowl of lime Jell-O. They discovered, by attaching the bowl to an EEG machine, that the motion of the set Jell-O has exactly the same waveforms as the human adult brain! See http://everything2.com/index.pl?node=Jell-o (accessed January 9, 2008).

7. Daniel J. DeNoon, "Rx for Diabetes: Laughter," (*WebMD Medical News*, May 28, 2003) http://diabetes.webmd.com/news/20030528/rx-for-diabetes-laughter (accessed March 6, 2008). Also see http://www.coolquiz.com/trivia/explain/docs/laughter.asp. Just for fun, check out http://www.youtube.com/watch?v=UjXi6X-moxE and http://www.youtube.com.watch?v=3ky-uo.

8. "Therapeutic Benefits of Laughter," (Holistic Online.com, n.d.) http://www.holisticonline.com/humor_therapy/humor_therapy_benefits.htm (accessed March 6, 2008).

9. For more information about Marilyn Sprague-Smith, visit http://www.miraclesmagicinc.com/threapeutic_laughter.htm.

10. Susan Phinney, "Humor has fans in medical circles," *Seattle Post Intelligencer,* March 14, 2006. http://seattlepi.nwsource.com/health/262840_laughter.14.html (accessed March 6, 2008).

11. For more information about Claiborne and The Simple Way, visit http://www.thesimpleway.org/index2.html.

12. Frederick Buechner, *Wishful Thinking: A Theological ABC* (New York, Harper and Row, 1973), n.p. http://www.academic.evergreen.edu/curricular/ireland200304/poetry/Neither_the_hair.htm (accessed April 14, 2008).

13. "Oprah Gail Winfrey," *Human Archives.org.* http://oprahwinfrey.humanarchives.org (accessed January 8, 2008).

14. Rev. Motlalepula Chubaku, "Going Up to the Mountain." *Peace Newsletter* (New York: The Syracuse Peace Council, Oct. 1985). www.peacecouncil.net/history/PNLs1981-90/PNL520-1985.pdf

15. Leonard Sweet, public teaching at a youth event.

16. "A New Creed," The United Church of Canada. http://www.united-church.ca/beliefs/creed (accessed March 6, 2008).

17. Gloria's Web site is located at www.viaCREATIVA.net.

18. Dr. Catherine Kapikian. Quoted in a class lecture, May 28, 2007, Wesley Theological Seminary, Washington D.C. Director, Henry Luce III Center for Arts and Religion.

19. Sweet, *AquaChurch*, 30.

20. William Shakespeare, *Hamlet*, Act 3, Scene 1 (New York: Washington Square Press, 2003).

21. http://www.archives.cnn.com/TRANSCRIPTS/0312/25/ltm.07.html (accessed April 14, 2008).

22. Sister Karol Jackowski, *Ten Fun Things to Do Before You Die* (Notre Dame, IN: Ave Maria Press, 1989), 12.

23. Thanks to Gloria Hughes for the testimony and the idea.

Chapter 9: Save

1. "The *rete mirabile*," *Britannica Online*, http://www.britannica.com/eb/article-51583 (accessed June 25, 2007).

2. Janet A. Simons, Donald B. Irwin, and Beverly A. Drinnien. "Maslow's Hierarchy of Needs" from *Psychology: The Search for Understanding* (New York: West Publishing Group, 1987).

3. "Rescue the Perishing," words by Frances Jane (Fanny) Crosby, 1870 (public domain). This hymn focuses on salvation from sin and eternal death. The opening line of its chorus, "Rescue the perishing, care for the dying," can be interpreted as a wider call for the active rescue of people in this life as well.

4. *Britannica Online*, http://www.britannica.com/eb/article-51583.

5. Ibid.

6. Ibid.

7. For more information, visit http://www.themeaslesinitiative.com. Also, Neil the Ethical Werewolf, "You Can Save More Lives Without Killing People,"http://www.ezraklein.typepad.com/blog/2006/06/you_can_save_mo.html.

8. The SCHIP program was a part of the five-year omnibus Farm Bill from 2003-2007. It was a successful, cost-effective partnership of the federal government and the states that insured millions of children, which came up for renewal in the 2007 legislature. All of the states endorsed it and both Democrats and Republicans overwhelming voted for it. President Bush vetoed it twice, scolding legislators for extravagant domestic spending.

9. For more wisdom from Einstein, visit http://www.thinkexist.com/quotes/Albert_Einstein/.

10. Joseph Cardinal Bernardin, "A Consistent Ethic of Life: An American-Catholic-Dialogue." Gannon Lecture, Fordham University: December 6, 1983. http://www.priestsforlife.org/magisterium/bernardinwade.html. For the complete lecture, go to www.paxjoliet.org/advocacy/bernardin/gannonlecture3.htm.

11. Jim Wallis, "An Open Letter to Chuck Colson," *SoJoMail*, February 24, 2005, http://www.sojo.net/index.cfm (accessed June 15, 2007).

12. *Pay It Forward*. Warner Brothers Pictures, 2000.

Forward: Being Human—Beyond the Porpoise-Given Life

1. Chris Hughes. *The NeoChurch.net Project* (Drew University, D.Min. project, 2004), 5.

2. Leonard Sweet, *Jesus Drives Me Crazy: Lose Your Mind, Find Your Soul* (Grand Rapids, MI: Zondervan, 2003), 76; Paul J. Achtemeirer, *Romans, Interpretation: A Bible Commentary for Teaching and Preaching* (Atlanta, GA: John Knox Press, 1985), 212.

3. Hughes, *The NeoChurch.net Project*, 5.

4. Ibid., 6.

5. Paul Tillich, *The Eternal Now* (New York: Charles Scribner's Sons, 1963).

6. Leonard Sweet. From a class lecture, January 18, 2002.

7. Alan Kay, quoted at http://www.smalltalk.org/alankay.html.

About the Author

Chris B. Hughes is a passionate post-postmodern prophet, pastor, multimedia communicator, manic metaphorager and amateur cyclist, cook, and dol-fan.

He is available for retreats, worship concerts, and ministry consultations. Visit his Web site at **www.chrisbhughes.net**.